LABORATORY
MANUAL

LABORATORY MANUAL

Daphne Norton
University of Georgia

CHEMISTRY
STRUCTURE AND PROPERTIES
Second Edition

NIVALDO J. TRO

 Pearson

Courseware Portfolio Management Director: Jeanne Zalesky
Executive Courseware Portfolio Manager: Terry Haugen
Product Marketer: Elizabeth Ellsworth
Managing Producer: Kristen Flathman
Courseware Director, Content Development: Jennifer Hart
Courseware Analyst: Coleen Morrison
Development Editor: Erin Mulligan
Portfolio Management Assistant: Sherican Kinosian
Rich Media Content Producer: Jackie Jacob
Content Prodcuers: Lisa Pierce, Laura Perry
Full-Service Vendor: codeMantra
Design Manager: Marilyn Perry
Cover and Interior Designer: Jeff Puda
Illustrators: Lachina
Rights & Permissions Project Manager: Ben Ferinni
Rights & Permissions Management: Cenveo Publisher Services
Photo Researcher: Eric Schrader
Manufacturing Buyer: Maura Zaldivar-Garcia
Cover Art: Quade Paul

ISBN-10: 0-134-61645-6; ISBN-13: 978-0-134-61645-2

This lab manual is dedicated to the hundreds of teaching assistants at Emory University who provided constructive feedback in the development of these experimental procedures. I offer a special *thank you* to Drs. Kevin O'Halloran, Ana West, and Teddy Huang for their individual contributions to the text.

Table of Contents

viii

The goal of this lab manual is to develop critical-thinking and problem-solving skills while fostering student engagement in real-world applications of chemistry.

Through real-world connections, students recognize the impact science has on their daily lives! The laboratory is a site for investigation and exploration. It lends the opportunity to visualize theories and evaluate principles of chemistry. It is a place for hands-on learning.

This laboratory manual provides a series of experiments written to correspond with an atoms-first approach. A significant challenge with this approach is the late introduction of stoichiometry. Many chemists are hesitant to adopt this format of teaching for fear of negatively impacting the laboratory. As a lab director, I was initially skeptical. However, students are unaware of the biases faculty have with regard to the sequence of course material. They will accept the challenge placed before them and with the appropriate support will succeed in mastering new skills.

This laboratory manual includes experiments specifically designed to connect with *Chemistry: Structures and Properties*. For example, students will evaluate the energy efficiency of a series of light bulbs in an experiment entitled *Energy & Electromagnetism: Irradiance Measurements*. This experiment builds on the principle that light can be viewed as a stream of particles called photons, and the energy of those photons can be measured. Other early experiments focus on experimental design and apply the scientific method to a laboratory setting. Exercises related to atomic structure and molecular bonding are provided before experiments dealing with synthesis and reactivity.

REAL-WORLD APPLICATIONS

Students become more engaged when they are presented with real-world applications that connect to their daily lives. This laboratory manual includes experiments that connect to household items such as Coca-Cola®, fertilizer, light bulbs, and aluminum cans. Yet the investigations are at a challenging level that prevents the exercise from becoming simple "kitchen chemistry." The problem-based nature of many experiments promotes critical-thinking skills and avoids a "cookbook" approach of following a series of steps in a procedure.

Experiments will expose students to recent advances in science as synthesizing *liquid crystals* and prepare *nickel nanowires*. These experiments provide a brief introduction to nanotechnology and materials chemistry. Students will prepare new materials found in electronic devices and discover the interface between chemistry and engineering.

EMPHASIS ON SUSTAINABILITY

Real-world applications would not be complete without including issues related to energy consumption and emissions. In the *Alternative Fuel Project*, students will research energy sources and write a proposal for a new transportation center on campus. As they learn about biodiesel and fuel cell technologies, they also recognize the interdisciplinary nature of these issues. Recycling is the focus of another experiment where students convert metal from an aluminum can into household material. Post-lab questions are also included to connect students to environmental problems for experiments such as calorimetry, energy, and freezing point depression. The leaf icon shown here will be used to show which problems carry an environmental theme.

APPLIED CHEMISTRY: SOLVING A PROBLEM

The labs in this manual promote critical thinking by placing the experiment in the context of a practical problem. Investigations place significance on data collection and analysis. For example, iron analysis requires the diagnosis and treatment of a patient. Students will analyze the amount of phosphoric acid in Coca-Cola®. In another experiment, they must determine what dyed foods might harm a friend who has a food allergy.

Some labs will require students to develop a method or experiment to gather information or solve a problem. However, the objectives are clearly outlined, and steps are in place where the student must discuss methods with the instructor before proceeding with the experiment. Pre-lab questions are written to guide the student through the reasoning process necessary to solve the problem presented and provide the necessary background for the experiments. The aim is to also guide the student through the decision making and calculations necessary for data analysis.

OFFERS FLEXIBILITY

This manual includes a sample of problem-based and traditional experiments to offer flexibility to the instructors. Some of the exercises are inquiry-driven, while others provide a straightforward method for introducing new laboratory techniques. Throughout the semester, the student will collaborate as part of a team and also work as an individual or with a lab partner. In some cases, as with the analysis of fertilizer, a group of four students will work together, but each individual team member will analyze a separate sample. They may rely on one another to practice techniques and perform calculations. However, each student will have his or her own set of data. This allows autonomy within a group setting.

Diagrams for techniques such as pipetting and vacuum filtration are placed in the text of the experiment, as I have found that students are less likely to flip to an Appendix to read instructions for a technique. The information is available as it becomes necessary for a procedure.

These experiments have all been vetted by thousands of students and hundreds of teaching assistants. Instructors may adapt the experiments to fit the students and facilities at their institution. I am aware of the challenges of teaching large enrollment classes and working with graduate teaching assistants. These experiments can easily be modified to include more or less inquiry.

Exploration at the beginning of the semester heightens student excitement for research. They realize that science can be messy and unpredictable and exciting!

Daphne Norton

CHEMISTRY LABORATORY SAFETY

- Absolutely no food, drink, or gum may be consumed in the laboratory. If you carry a water bottle, it should be stored inside your backpack. You may not have your water bottle out on the bench top.

- All cell phones, pagers, and other electronic devices must be turned off before entering the lab.

- You must wear eye protection at all times while in the laboratory! Goggles or safety glasses must be worn over regular eyeglasses.

- Sandals may not be worn in the lab. You should wear closed-toed shoes. You may not wear flip flops or clogs.

- The appropriate laboratory attire includes long pants and a shirt that completely covers your body. No shorts, tank tops, halter tops, or cut-off shirts are allowed. Avoid synthetic fabrics such as nylon, rayon, and spandex. These fabrics dissolve in common solvents.

- The lab is equipped with a safety shower and eyewash.

CHEMICAL SAFETY

- Caution should be used when handling chemicals.

- Material Safety Data Sheets are available for each chemical in the lab. These sheets provide detailed information about chemical hazards.

- Never return chemicals to the reagent bottles.

- Pour solvent into a beaker and pipet from the beaker. Do not pipet directly from the solvent bottle.

- If you spill any acid, base, or other chemical, contact your instructor immediately.

- Be smarter. Add acid to water. (Do not add water to acid.)

- Thoroughly wash your skin of any chemical contaminants.

- If any chemical gets in your eye, you must contact your instructor immediately and proceed to the eyewash station. It may be helpful to have another student assist you in walking to the eyewash or alerting your instructor. Flush your eye with water for a minimum of 20 minutes.

- If you encounter a chemical splash or a fire, you must immediately go to the safety shower. Alert your instructor who will assist you during the emergency. It is critical that you remove clothes that are soaked in chemicals.

- Dispose of chemical waste in the appropriate waste container. Some waste containers may be located in the hood. Carefully read the label and do not mix waste materials as this may result in a dangerous chemical reaction. DO NOT pour any chemicals down the drain without permission from your instructor.

GENERAL EQUIPMENT USAGE

- In the event of glassware breakage, contact your instructor who will assist you in cleaning up the broken glassware. All broken glassware should be placed in the broken glassware receptacle.

- Keep the work area free of debris. At the end of the period, no trash or chemicals should be left on the bench tops.

- Any equipment taken from the cabinets should be returned to its appropriate place.

- Hotplates should be close to room temperature before returning to the shelf.

- Cords should be wrapped around the hotplates and stir plates before returning these items to the shelf.

- Check that the gas outlets and water faucets have been turned off after use.

- You must clean any spills and trash before leaving the lab. Your instructor may check your work station before you leave.

Liquid Crystals | 1

bloomua/Shutterstock

IN-BETWEEN PHASES OF MATTER: LIQUID CRYSTALS

Materials Chemistry can be defined as the branch of chemistry aimed at the preparation, characterization, and understanding of substances that have some specific useful function. The discipline can be divided into categories:

- Preparation/synthesis (How are materials made?)
- Structure (How are they put together?)
- Characterization (How do they behave?)
- Applications (How can we use them?)

Materials Chemistry largely involves the study of condensed phases (solids, liquids, and polymers) and interfaces between different phases. Because many of these materials have direct technological applications, Materials Chemistry has a strong link between basic science and many existing and newly emerging technologies. For this experiment, we will explore the interface between liquids and solids in the preparation of cholesteryl ester liquid crystals. We will later investigate the behaviors of these same materials.

Liquid crystals were first discovered in 1888 by Friedrich Reinitzer, an Austrian botanist and chemist. He noticed what appeared to be two separate phase changes when he was studying cholesterol extracted from carrots.[1] A liquid crystal is an organic compound with properties that appear to be fluid and crystalline simultaneously. Liquid crystals have unique optical properties that respond to changes in thermal, electric, and magnetic fields. Although these liquids flow, they are more "ordered" than normal liquids, as seen in Figure 1.1. They can still dissolve other materials and possess surface tension.

1

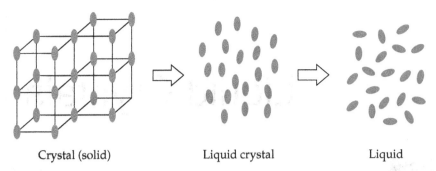

Crystal (solid)   Liquid crystal   Liquid

FIGURE 1.1 A crystalline solid is highly ordered. There is less order in a liquid crystal and even less in a liquid.
Reprinted by permission of Chelix Tech Corp.

There are three types of liquid crystal: nematic, smectic, and cholesteric. The molecules align along a "director." The long axes of the molecules tend to align in the direction shown in Figure 1.2.

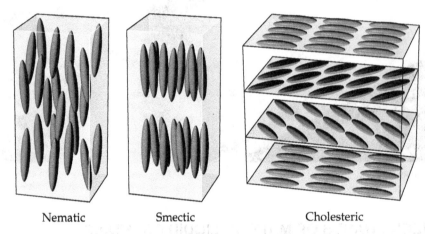

Nematic   Smectic   Cholesteric

FIGURE 1.2 The position of the layers in the cholesteric liquid crystal shift or twist along a helical axis.
Based on Brown et al., Chemistry: The Central Science 13e © 2015 Pearson Education, Inc.

The 1991 Nobel Prize in Physics was awarded to Pierre-Gilles de Gennes "for discovering that methods developed for studying order phenomena in simple systems can be generalized to more complex forms of matter, in particular to liquid crystals and polymers."[2] De Gennes made his chief contribution to the field of liquid crystals when he explained what is termed as anomalous light scattering from nematic liquid crystals. This light scattering depends in a complicated manner on fluctuations in the orientational order. Another important contribution was his description of the behavior of liquid crystals when a weak alternating electric field is applied. Materials scientists have collaborated with engineers to create LCDs, or liquid crystal displays, found in everyday products such as laptops, digital watches, and flat-screen televisions.

Working with a partner, you will prepare a *cholesteric liquid crystal* with a transition at a specific temperature range. As seen in Figure 1.2, a cholesteric liquid crystal is aligned in sheets that are rotated with respect to one another. This gives a rotation of light. This type of liquid crystal is typically composed of molecules containing a chiral center.[3] The term *chiral* refers to a molecule that has a nonsuperimposable mirror image. This often arises when central carbon atoms in the molecule are asymmetric. In this case, all of the groups attached to the carbon atom are different. In liquid crystals, chirality produces intermolecular forces that favor alignment between molecules at a slight angle to one another. A cholesteric liquid crystal is formed by layers. It is said to be twisted because the molecules align in layers and that alignment rotates slightly between the layers, eventually bringing the molecules back into the original orientation.

TABLE 1 Individual Compounds and Their Ratios for the Cholesteryl Liquid Crystal

Cholesteryl oleyl carbonate	Cholesteryl pelargonate	Cholesteryl benzoate	Transition range, degrees C
0.65 g	0.25 g	0.10 g	17–23
0.70 g	0.10 g	0.20 g	20–25
0.45 g	0.45 g	0.10 g	26.5–30.5
0.43 g	0.47 g	0.10 g	29–32
0.44 g	0.46 g	0.10 g	30–33
0.42 g	0.48 g	0.10 g	31–34
0.40 g	0.50 g	0.10 g	32–35
0.38 g	0.52 g	0.10 g	33–36
0.36 g	0.54 g	0.10 g	34–37
0.34 g	0.56 g	0.10 g	35–38
0.32 g	0.58 g	0.10 g	36–39
0.30 g	0.60 g	0.10 g	37–40

Preparing a Liquid Crystal

Select one of the preparations[4–6] shown in Table 1, or one may be assigned by your instructor. Tare the mass of a weigh boat or piece of weigh paper. Measure the exact amount shown for each of the reagents and carefully transfer the solid reagents into the vial. All three reagents will be added to the same vial. Use a separate weigh boat for each measurement. Be careful to not contaminate the reagent bottles!

Use a heat gun to melt the three reagents in the vial. The solids will melt into a liquid. The liquid crystal will begin to solidify as it cools to room temperature. Try to evenly coat the inside walls of the vial by rolling the vial on the bench top during the heating process. Use a pair of tongs to roll and pick up the vial. *Caution: The glass becomes very hot during the melting process.* Record your observations throughout the preparation.

Designing an Experiment to Test Another Group's Liquid Crystal

Next, you will receive a vial containing a liquid crystal prepared by another team in the lab. Design a method to determine the active temperature range for this liquid crystal. Discuss your method with your instructor before testing the sample.

Useful websites where you can learn more about liquid crystals include:

- http://plc.cwru.edu/tutorial/enhanced/files/lc/intro.htm
- http://www.nobelprize.org/educational/physics/liquid_crystals/
- http://plc.cwru.edu/tutorial/enhanced/files/textbook.htm

REFERENCES

1. Lehmann, O. (1889). "Über fliessende Krystalle." *Zeitschrift für Physikalische Chemie* 4: 462–72.
2. "Press Release: The 1991 Nobel Prize in Physics." Nobelprize.org. Nobel Media AB 2013. Web. 2 Oct 2013.
3. http://www.doitpoms.ac.uk/tlplib/anisotropy/liquidcrystals.php
4. Brown, G. H., and J. J. Wolken. 1979. *Liquid Crystals and Biological Systems.* New York: Academic Press: 165–167.
5. Elser, W., and R. D. Ennulat. 1976. *Adv. Liq. Cryst.* 2 (73).
6. Lisensky, George, and Elizabeth Boatman. 2005. *Journal of Chemical Education* 82 (9): 1360A.

PRE-LAB QUESTIONS | EXPERIMENT
Liquid Crystals | 1

Some of these questions will require you to gather information from outside sources. Much valuable information is found online; however, critical evaluation of each website is necessary. Content presented by national laboratories and university research sites is a good starting point. Answer the questions in your own words, including a reference if appropriate.

1. Most liquid crystals exhibit polymorphism. Explain *polymorphism*.

2. Liquid crystals are described as anisotropic. Define *anisotropy*.

3. Describe the continuum of liquid—liquid crystal—solid. Discuss the differences in terms of physical properties.

4. Define order parameter, S, and relate this value to the structure of each substance: liquid, liquid crystal, and crystalline solid.

5. Most liquid crystal molecules are rod-shaped and are broadly categorized as either thermotropic or lyotropic. Explain the two categories.

6. Cite two to three practical applications of a lyotropic liquid crystal.

Name _____ Date _____

Instructor _____

REPORT SHEET | EXPERIMENT

Liquid Crystals | 1

Using Table 1 of your lab manual, prepare a liquid crystal. You may pick from any one of the 12 liquid crystals shown in the table or your instructor may assign one to you. *Do not discuss the active temperature range with other groups in the room!*

Describe your observations as you prepare the liquid crystal. Explain possible draw-backs of this procedure. Propose a more efficient method of preparation.

This week you prepared a cholesteric-nematic liquid crystal. Explain the properties of this class of liquid crystals. What does it mean for a molecule to be chiral?

DESIGNING AN EXPERIMENT TO TEST ANOTHER GROUP'S LIQUID CRYSTAL

Explain how you will test the sample. Be specific. *Describe your method to your instructor before beginning to test the sample.*

Outline your method below.

At what temperature do you see a color change? Do you see multiple transitions? Describe your observations.

Were you successful in identifying the active temperature? Explain possible sources of error. Comment on limitations of your method.

Atomic Emission Spectra—Comparing Experimental Results to Bohr's Theoretical Model | 2

The goal of this exercise is to consider the evolution of scientific understanding and science's attempt to explain natural phenomena. We will evaluate the models used to describe the internal interactions of an atom. We will be testing a model to move toward a better understanding of electronic energy.

Problem:

Scientists could not understand why the emission spectrum for hydrogen was not a continuous spectrum. This phenomenon contradicted Rutherford's nuclear model.

Postulation:

Neils Bohr proposed a model for the hydrogen atom that stated that the electron was said to orbit the nucleus in certain "allowed" orbits.

What did they know back in 1913?

Experimental findings:

When sunlight was refracted through a prism, a rainbow of colors was observed. Scientists referred to this as a *continuous spectrum*. This demonstrated that electromagnetic radiation behaves like a wave, as shown in Figure 2.1.

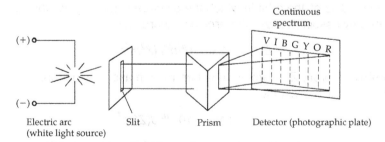

FIGURE 2.1 When white light is passed through a prism, the component wavelengths create a continuous spectrum in the colors of the rainbow.
From Tro, Laboratory Experiments for Chemistry: The Central Science 12e. © Pearson Education, Inc.

When an electron discharge was passed through a sample of hydrogen gas, an emission spectrum was observed that had distinct lines in the visible region, as we can see in Figure 2.2.

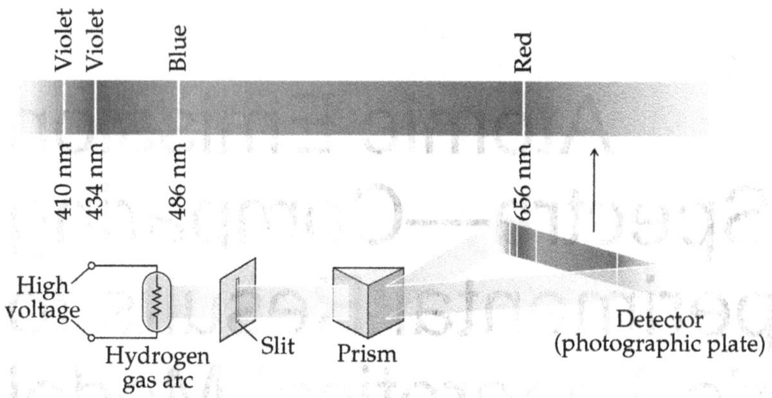

FIGURE 2.2 The emission spectrum of hydrogen, also called a line spectrum.
From Tro, Laboratory Manual for Chemistry: A Molecular Approach 3e. © Pearson Education, Inc.

In 1885, Joseph Balmer predicted that as n increased, the frequencies would become more closely spaced. When photographic paper became available, Johann Rydberg expanded this understanding to account for transitions that emit lines in the UV and IR regions.

Principles of Bohr's Theoretical Model

According to Bohr, electrons moved in discrete orbits or energy levels. At the lowest energy level, the electron is said to be in the ground state, $n = 1$. When excited by external forces, the electron moves to a higher-energy level referred to as an excited state.

When an electron in a high-energy level falls to a lower-energy level, the electron emits radiation that is equal to the difference in the two energy levels. The energy is emitted as a photon.

Max Planck showed that the energy of this photon is directly related to the frequency of the wave of motion of the photon, where h represents Planck's constant defined as $6.626e^{-34}$ J·s and c is the speed of light at $3.0e^{8}$ m/s.

$$E = \frac{hc}{\lambda}$$

Bohr stated that orbits were allowed only if the angular momentum of the electron in that orbit was divisible by $(h/2\pi)$. Angular momentum is the product of the mass and velocity of the electron times the radius of the orbit in which the electron is moving. As the radius of the orbit increases, the energy associated with the orbit also increases.

$$E_n = -(2\pi\, me^4/h^2)\,(Z^2/n^2)$$

where n = number of orbit, Z = atomic number, m = mass of electron, e = charge on the electron, and h = Planck's constant

$$E_n = -2.18 \times 10^{-18}\ J(Z^2/n^2)$$

The difference in energy levels can easily be determined from this equation where n = number of orbit or energy level. The equation for hydrogen is shown as:

$$\Delta E = E_{final} - E_{initial} = -2.18 \times 10^{-18}\ J(1/n_{final}^2 - 1/n_{initial}^2)$$

Bohr also predicted the wavelength, λ, associated with these energy transitions. For example, the transition for an electron in the second energy level dropping to the ground state is shown as:

$$\lambda_{2 \to 1} = hc/E_2 - E_1$$

PROCEDURE

Practical Exercise Part I: Comparing Theory and Experimental Observations

We will compare Bohr's theoretical predictions with experimental observations. First, we will observe a hydrogen emission spectrum. Then, we will calculate the energy levels associated with the first six energy levels for hydrogen and predict the wavelengths associated with various transitions.

Bohr's model can only be applied to a one-electron system. The model works nicely for hydrogen. However, it does not work for a multielectron system.

Our observations will be made with a handheld spectroscope. Your instructor will demonstrate proper use of the device. The spectroscope may require calibration using a mercury lamp.

1.   First, observe the line spectrum for hydrogen gas using the handheld spectroscope. Record your observations in the appropriate column on the Report Sheet. (Experimental λ)

2.   Calculate the energy level for the first six orbits of the hydrogen atom. Report energy values in joules.

3.   Find the energy difference (in joules) for all possible transitions between the first six orbits and record those values in the table. Using these energy differences, calculate the wavelength (in nm) associated with the energy difference. (1 nanometer = 10^{-9} m)

   An example is provided in the top left corner of the table. This is the energy difference between $n = 6$ and $n = 1$ ($E_6 - E_1$).

4.   Label the transitions associated with each wavelength.

Part II: Observing Emission Spectra for Group 1A and Group 2A Metal Ions

Hydrogen exists naturally in the gas state. This requires us to observe the element in a gas tube. Metals in Group 1A and Group 2A form ions in aqueous solutions. A solution of these metal ions can be introduced to a Bunsen burner flame. The flame excites the electrons to a higher energy level. When the electrons return to their initial states, energy is emitted (see Figure 2.3).

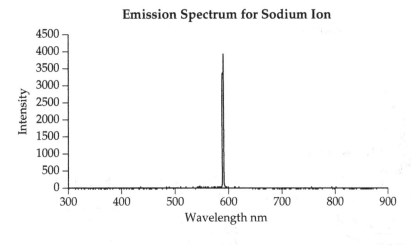

Emission Spectrum for Sodium Ion

FIGURE 2.3 An emission spectrum of a 1.0 M solution of NaCl. The sodium ions were excited when introduced to a Bunsen burner flame.

When they become excited, common metallic ions emit light very strongly in the visible region. The emission is so strong that the elements may often be visually detected by the overall color they impart to a flame. A spectrometer may be used to plot an emission spectrum. An emission spectrum plots the intensity of the emission versus wavelength as shown in Figure 2.3. The energy associated with the electronic transition can be calculated if the wavelength associated with the emission can be observed and measured. The wavelength may be detected by a handheld spectroscope or a spectrophotometer.

The wavelength observed in Figure 2.3 can be used to calculate the energy associated with the transition.

589.3 nm $\quad E = \dfrac{hc}{\lambda} = \dfrac{(6.63 \times 10^{-34} \text{ J/s})(3.00 \times 10^{17} \text{ nm/s})}{589.3 \text{ nm}} = 3.38 \times 10^{-19} \text{ J/atom}$

$$E = \dfrac{3.38 \times 10^{-19} \text{ J}}{\text{atom}} \times \dfrac{1 \text{ kJ}}{1000 \text{ kJ}} \times \dfrac{6.02 \times 10^{23} \text{ atoms}}{1 \text{ mol}} = 203 \text{ kJ/mol}$$

The emission spectrum of each element is unique. Although several metals may produce the same color flame, their emission spectra are specific to the wavelengths associated with energy transitions specific to each element. Use the following steps to work with a partner to observe the emission spectra for Li, Na, K, Ca, Sr, and Ba ions. Safety glasses should be worn throughout the experiment.

1. Place ~1 mL of the 1.0 M aqueous sodium chloride solution onto a watch glass.

2. Obtain a nichrome wire. The nichrome wire must be cleaned before it can be used to introduce samples into the flame. Clean the wire by dipping it in 6 M HCl and swirling it in the solution for 15–20 seconds. *Caution: If acid makes contact with your skin, immediately wash the area with water and contact your instructor.*

3. Rinse the wire with distilled water. Wash it thoroughly to remove residual acid.

4. Carefully light the Bunsen burner.

5. Use the clean wire to introduce the solution into the flame. You will tilt the watch glass so that it leans toward the air intake of the Bunsen burner shown in Figure 2.4. Tap the solution with the coiled wire to splatter solution into the air intake. The solution will be sucked up into the flame. (You may also use an atomizer to introduce the solution. Your instructor will provide specific instructions.)

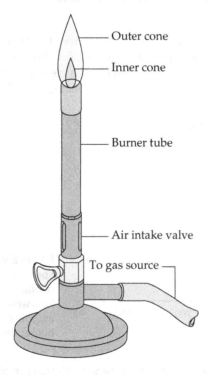
— Outer cone
— Inner cone
— Burner tube
— Air intake valve
To gas source —

FIGURE 2.4 The air intake is at the bottom of the burner tube.
From Heasley, Victor L.; Christensen, Val J.; Heasley, Gene E., Chemistry and Life in the Laboratory: Experiments, 6th edition © 2013 Pearson Education, Inc., pg. 34

6. As the solution reaches the flame, your lab partner should observe the emission using a handheld spectroscope or an emission spectrum should be recorded using a fiber-optic sensor connected to a spectrophotometer. (Again, your instructor will provide specific instructions.)

7. Repeat the previous steps for the other metal ion solutions and your unknown. Discard the acid and solutions in the appropriate waste containers.

Name ____________________ Date ____________________

Instructor ____________________

PRE-LAB QUESTIONS | EXPERIMENT

Atomic Emission Spectra—Comparing Experimental Results to Bohr's Theoretical Model

2

1. What is a model? How do scientists use models?

2. Explain Rutherford's nuclear model of an atom.

3. What is a continuous spectrum?

4. What is an emission spectrum?

5. Neils Bohr derived an equation for calculating energy levels of an atom:

$$E_n = -2.18 \times 10^{-18} \, J(Z^2/n^2)$$

Why does the equation have a negative sign, resulting in a negative energy value?

6. It will be necessary to calibrate the spectroscope. What does it mean to calibrate something? Explain calibration and give an example of something in a household or office that may need to be calibrated from time to time. Explain why the object or device needs calibration.

Name ______________________ Date ______________________

Instructor ______________________

REPORT SHEET | EXPERIMENT

Atomic Emission Spectra— Comparing Experimental Results to Bohr's Theoretical Model

2

Hydrogen Lamp/Emission Spectrum

Color	Experimental λ (nm)*	Calculated λ (nm)	Energy (kJ/mol)	Transition
red				
turquoise				
violet				
purple				

* If you do not physically observe the line, write NOT OBSERVED in the column for the wavelength.

Theoretical Calculations of the Energies of the First Six Orbits of the Hydrogen Atom

Orbit Predicted Energy	Joules
n_1	
n_2	
n_3	
n_4	
n_5	
n_6	

Observed λ in nm	Probable Transition
97.2	
102.6	
121.6	
433.4	
486.1	
656.2	
1281.8	
1875.1	
4050.0	

n_{low}		n_{high} 6	5	4	3	2
1	ΔE	2.118×10^{-18} J				
	λ	93.7 nm				
2	ΔE					
	λ					
3	ΔE					
	λ					
4	ΔE					
	λ					
5	ΔE					
	λ					

1. State one limitation of Bohr's model of the atom.

2. Why don't we observe the transition from $n_2 \rightarrow n_1$ for the hydrogen atom in the line spectrum? Be specific.

Part II Flame Emission Observations

Metal Ion	Flame Color	Most intense lines (wavelengths) observed
Li		
Na		
K		
Ca		
Sr		
Ba		

Unknown ID	Flame Color	Most intense lines (wavelengths) observed

Identity of Single Salt Unknown _____

Justify the identification of your unknown.

Calculate the energy associated with the wavelength(s) present in your single-ion unknown.

Identity of Mixed Salt Unknown _____

Justify the identification of your unknown.

Energy & Electromagnetism— Irradiance Measurements | 3

The invention of the light bulb and the creation of a system for the distribution of electricity revolutionized the lives of American citizens. Prior to the invention of the light bulb, individuals burned oil lamps, lit candles, and were dependent on natural light sources. Most activities took place during the day with the aid of natural sunlight. Sunlight is a collection of electromagnetic radiation emitted from the sun. The sun gives off ultraviolet, visible, and infrared photons as shown in the solar irradiance spectrum in Figure 3.1. This spectrum is similar to that from a blackbody radiator at ~5500 K, which is approximately the temperature on the surface of the sun. A blackbody radiator is an opaque and nonreflective body that is in thermodynamic equilibrium with its environment when held at a constant temperature. When sunlight passes through the atmosphere, 11.5% of the solar energy is absorbed by H_2O, O_2, and CO_2 molecules. It is estimated that the solar flux reaching the earth's surface consists of 5.6% ultraviolet (< 400 nm), 43.6% visible (400–700 nm), and 50.8% infrared (> 780 nm) irradiation in terms of power.[1] Power is energy divided by time.

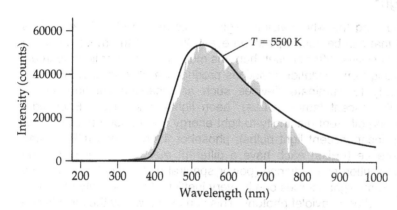

FIGURE 3.1 The solar irradiance spectrum at the top of the atmosphere is shown in the shaded region. The dark line represents a spectrum characteristic of a blackbody radiator. This figure is based on reference spectra from the American Society for Testing and Materials (ASTM).[2]
Based on Knight, Physics for Scientists and Engineers: A Strategic Approach with Modern Physics, 2nd edition, © 2008 Pearson Education, Inc., pg. 1200 and data from the author.

Spectral View of Magnetism

A spectrophotometer is a device for measuring light intensity as a function of the wavelength of the light source. We can use a spectrophotometer to measure the energy that reaches a given surface. Irradiance describes the power per unit area of each wavelength of electromagnetic radiation at a given surface. The unit of power is the watt, W (one W = one J/s). Therefore, irradiance is often given in W/m². Absolute irradiance is the measure of light in absolute terms, and this is the rating we see on light bulbs (40 W, 60 W, etc.). This rating states the total radiation power provided by the bulb but gives no information about the wavelength of the emitted radiation.

Relative irradiance is a measurement of light relative to the known color temperature of a blackbody light source. It compares the fraction of energy emitted by the sample to the energy collected from a lamp with a blackbody energy distribution (normalized to 1 at the energy maximum). Relative irradiance allows the user to determine whether there is more light at one wavelength than another but does not provide power in absolute terms. Common applications include characterizing the light output of light-emitting diodes (LEDs), incandescent lamps, and other radiant energy sources such as sunlight. We will measure the relative irradiance for a series of light sources and correct the spectra relative to a tungsten lamp at 3100K. A tungsten lamp behaves similarly to a blackbody radiator. It will serve as our reference. See Figure 3.2.

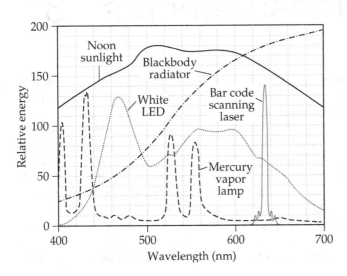

FIGURE 3.2 By comparing our collected relative irradiance spectra with the solar radiation spectrum, we can suggest which light source most closely matches the display in natural sunlight. The figure to the left provides spectra for a variety of common light sources including a blackbody radiator.[3] Sunlight absorbs light across the entire visible spectrum, while a bar code scanning laser reveals a sharp peak at 633 nm because its operating wavelength is in the red region.

Spectra from Common Sources of Light, from "Sources of Visible Light," Microscopy Resource Center, available at http://www.olympusmicro.com/primer/lightandcolor/ lightsourcesintro.html. Reprinted by permission of Florida State University.

Artificial Sources of Light

Sunlight contains photons spanning the whole visible region. This mixing of different colors results in a natural white light that can be seen by human eyes. Sunlight is the most important illumination source for human activities. After sunset, humans must rely on other light sources. For example, wood fires resulting from oxidation reactions produced a glow in ancient times. Currently, we rely on electricity to illuminate devices such as incandescent light bulbs, fluorescent lamps, compact fluorescent lamps (CFLs), neon lights, and LEDs. Each light source gives off photons following different electricity-to-light energy conversion principles. For example, blackbody radiation (incandescent light bulbs), phosphor fluorescence (CFLs), and semiconductor electroluminescence (LEDs) each have a different mechanism for producing light. Therefore, the power distribution of irradiance (power spectral distribution) can be very different for each light source. Some light sources can be very inefficient in electricity utilization due to the emission of infrared and/or ultraviolet photons. This can be shown by their irradiance spectra. An efficient light bulb or lamp should emit only visible photons.

Quality and Efficiency

The lighting efficiency of a source of illumination can be characterized by luminous flux (lumen per watt, lm/W), which can be adjusted to reflect the varying sensitivity of the human eye. Conversely, the overall electricity-to-electromagnetic radiation conversion efficiency is based on radiant flux (the total power of emitting light). Luminous flux is lumen per watt of input electricity and adjusted to reflect the varying sensitivity of the human eye. Therefore, the lighting efficiency is determined by two parameters: the electricity-to-electromagnetic radiation efficiency and the relative visibility of electromagnetic radiation. Traditionally, the maximum 683 lm/W for monochromatic green light (555 nm, peak sensitivity of human eyes) is defined as 100% luminous efficiency or lighting efficiency.[1,4] Other light sources are compared to this value for defining their light efficiency.

Analysis of Light Sources

In the real-life application, cost and life span are two additional factors that must be considered when installing light sources. Another consideration is the strength of radiation reaching the object. Energy in the UV region of the spectrum is particularly harmful. Special filters have been designed to block UV rays from being emitted. As we evaluate each light source, we should consider the energy emitted in the UV region along with the energy efficiency of each source. We will evaluate four common household light sources: light-emitting diodes (LEDs), compact fluorescent lamps (CFLs), fluorescent lamps, and incandescent light bulb. In this experiment, you will measure the power distribution spectra of several light bulbs and will evaluate each bulb to determine the percentage of emission in the UV, visible, and IR regions. You will also evaluate the temperature profile of the bulb and consider how much energy is lost as heat.

EXPERIMENTAL PROCEDURE

You will work in groups of four to measure the temperature of the bulbs provided by your instructor. Sample bulbs include the incandescent light bulb (60 W, Philips), compact fluorescent bulbs (13 W, Philips), fluorescent lamp in the lab, and light-emitting diode (12.5 W, Philips). You can consider the efficiency of the bulb by evaluating the temperature curves over time.

Caution: Never touch the illuminating incandescent light bulb. The temperature is higher than that of boiling water. Treat all bulbs carefully, as the bulbs and equipment are very fragile.

Temperature Measurements

1. Record the room temperature.
2. Insert the LED bulb into the lamp. (The lamp should be off.) Place the tip of the thermometer in touch with the LED bulb.
3. Start the timer as you turn on the lamp. Record the temperature at 30-second intervals for a total of 5 minutes. The tip of the thermometer should remain in contact with the bulb. (Do not touch the metal part of the temperature probe after it has made contact with the bulb as the metal probe will become very hot.)
4. Turn off the lamp. Replace the LED with another light bulb. (Wear cloth gloves when holding the bulbs.)
5. Repeat steps 2, 3, and 4 to record temperature measurements of the CFL and incandescent light bulbs.
6. Use Excel to plot the "temperature versus time" for each light source.

Irradiance Measurements

You will work in groups of four and use the USB4000 miniature fiber-optic spectrometer (Ocean Optics) to record irradiance measurements for the four bulbs provided by your instructor.

1. Turn off the overhead light in the room. Close the blinds to remove sources of natural light.
2. Set up the LED bulb. Measure and record the LED spectrum. Adjust the iris to make sure the spectrum measures a maximum 60000 without exceeding that value.
3. Your instructor will provide guidelines for copying your data to an Excel spreadsheet. When copying data to the spreadsheet, make sure you follow the instructions exactly. Data analysis by Excel is cell-position sensitive.
4. Turn off the LED. Measure and record the background signal. Copy these data to the spreadsheet.
5. Wear gloves and remove the LED light bulb. Replace the bulb with another one for the next measurement.

6.   Repeat steps 2–4 for measurements on the compact fluorescent lamp and incandescent light bulb. Make sure you copy the data to the right place.

7.   Turn on the room light and record the spectrum of the light from the fluorescent lamp in the ceiling fixture overhead. Make sure that the table lamp does not block the light from the fluorescent lamp. Adjust the iris to make sure the maximum of the spectrum is close to, but does not exceed, 60000. Record the data.

REFERENCES

1. Energy Efficiency & Renewable Energy (DOE), "Using Light-Emitting Diodes," http://www1.eere.energy.gov/buildings/ssl/efficacy.html

2. http://rredc.nrel.gov/solar/spectra/am1.5/

3. http://www.olympusmicro.com/primer/lightandcolor/lightsourcesintro.html

4. http://apps1.eere.energy.gov/buildings/publications/pdfs/ssl/report_led_november_2002a_1.pdf. Light Emitting Diodes (LEDs) for General Illumination, "An OIDA Technology Roadmap Update 2002."

Name _____ Date _____

Instructor _____

PRE-LAB QUESTIONS | EXPERIMENT

Energy & Electromagnetism—
Irradiance Measurements | 3

1. Sketch the electromagnetic spectrum, and label ultraviolet, visible, and infrared radiation.

 2. People often misunderstand the term radiation. It can have a negative connotation. Specifically consider ultraviolet radiation and cite two beneficial and two harmful effects of ultraviolet radiation.

   (a) Beneficial effects

   (b) Harmful effects

3. List two applications of infrared light.

4. What is a blackbody radiator?

5. Explain the mechanism by which an incandescent bulb emits light.

6. Define the following terms:
   lumen:

   flux:

7. Define and explain the difference between radiant flux and luminous flux.

8. Define and explain the difference between the terms watt and wattage.

Name _____ Date _____

Instructor _____

REPORT SHEET | EXPERIMENT

Energy & Electromagnetism— Irradiance Measurements | 3

Spectra

Attach a copy of the spectrum displaying the illuminating power distribution for each bulb.

Data Tables

TABLE 1: Illuminating Power Distribution (Excel file: spreadsheet "Results")

Light sources	Ultraviolet 200–400 nm	Visible 400–700 nm	Infrared 700–850 nm
LED			
CFL			
Incandescent			
Fluorescence			

TABLE 2. Temperature

Room temperature: _____°C   or   _____°F

Light sources	°C	°F
LED		
CFL		
Incandescent		

1. Explain the mechanism by which a fluorescent bulb emits light. Diagrams may be helpful.

2. Why are LED lights called "cold" lamps?

3. The spectrometer used has a spectral range from 200 nm to 850 nm. Photons with wavelengths longer than 850 nm will not be detected. Therefore, the spectrum for incandescent light bulbs is incomplete. So, is the measured ratio of infrared power overestimated, correctly estimated, or underestimated?

4. Assume 40% of the input electricity power is wasted as heat in an incandescent light bulb. (This value is higher in real life, but we will use this estimate.) If you have 10 incandescent bulbs in your apartment, calculate the energy wasted as heat per week. (Each bulb is 75 W (J/s), and your lighting duration is 6 hours per day.)

5. Combustion of 1.0 kg of carbon (graphite) will release 3.3×10^7 J of heat. How many grams of graphite will you need to burn to equal the heat wasted in Question 4?

6. Consider the data provided below (prices, input powers, and lifetimes) to determine the best investment.
   Incandescent light bulb: $0.35, 60 W, 1000 hours
   Fluorescent lamp: $1.25, 32 W, 24000 hours
   CFL: $2.1, 14 W, 10000 hours
   LED: $50, 10.5 W, 35000 hours
   Price of electricity: $0.12/kilowatthour ~ $0.12 per 3.6×10^6 J

7. Explain the mechanism by which a fluorescent bulb emits light. Diagrams may be helpful.

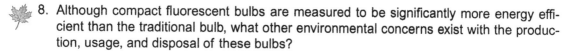 8. Although compact fluorescent bulbs are measured to be significantly more energy efficient than the traditional bulb, what other environmental concerns exist with the production, usage, and disposal of these bulbs?

9. What environmental concerns are associated with LEDs?

Structure of Molecules | 4

VSEPR Theory and Molecular Modeling with Ball-and-Stick Models

Many of the physical and chemical properties of molecules depend on the types of atoms in the molecule and the bonds that connect them. These properties are also affected by the way the atoms are arranged in space or the shape of the molecule. Therefore, the structural study of a chemical species (molecule, ion) is a very important field. Chemists can predict the shapes of many molecules from the chemical formula by drawing Lewis dot formulas and applying **Valence Shell Electron Pair Repulsion (VSEPR) theory**. VSEPR theory states that *electron pairs in a valence shell of an atom repel other electron pairs, and they minimize repulsion by making the angles between them as large as possible.*

In general, the following types of molecular shapes will occur based on the number of electron pairs around the central atom and the number of atoms that are bonded to the central atom. We will define the number of electron-dense regions surrounding the central atom as the steric number.

Summary of Molecular Structures and Polarity from the VSEPR Theory

Steric #	Electronic Geometry (Orbital Type)	Bonding Pairs	Molecular Shape	Point of Symmetry	Examples
2	Line (sp)	2	Line	Yes	$BeCl_2$
3	Trigonal Planar (sp^2)	2	Bent	No	SO_2
		3	Trigonal planar	Yes	BF_3
4	Tetrahedron (sp^3)	1	Line	No	HF
		2	Bent	No	SCl_2
		3	Trigonal pyramid	No	PCl_3, H_3O^+
		4	Tetrahedron	Yes	$CHCl_3$
5	Trigonal Bipyramid (sp^3d)	2	Linear	Yes	XeF_2
		3	T-shape	No	ClF_3
		4	Seesaw	No	SF_4
		5	Trigonal bipyramid	Yes	PF_5
6	Octahedron (sp^3d^2)	4	Square plane	Yes	XeF_4
		5	Square pyramid	No	BrF_5
		6	Octahedron	Yes	SeF_6

Classification of Simple Molecules/Ions and Their Shapes

(1) AB Type:	*always linear*	Examples: HCl, CO, OH$^-$
(2) AB$_2$ Type:	*linear or bent*	CO$_2$, H$_2$O, SO$_2$
(3) AB$_3$ Type:	*planar or pyramidal*	SO$_3$, CO$_3{}^{2-}$, NH$_3$
(4) AB$_4$ Type:	*planar, tetrahedral, or other*	CH$_4$, XeF$_4$, SCl$_4$
(5) AB$_5$ Type:	*pyramidal or bipyramidal*	PCl$_5$, IF$_5$
(6) AB$_6$ Type:	*octahedral*	SF$_6$, SnCl$_6{}^{2-}$

Electron Arrangements with Minimum Repulsion

# Areas of e$^-$ Density	Electronic Geometry	Arrangement
2	Linear	
3	Trigonal Planar	
4	Tetrahedral	
5	Trigonal Bipyramidal	90°
6	Octahedral	

Adapted from Experiments in General Chemistry Laboratory Manual, by Daphne Norton, Hayden McNeil Publishing.

AN EXERCISE WITH AMMONIA, NH₃

1. In the Lewis electron dot formula, the N atom must be placed at the center with four pairs of electrons around it. Each of the three H atoms is bonded by a pair of electrons (single bond) to nitrogen. One pair of electrons surrounding N is not used in the bonding.

2. The electron dot formula yields four electron-dense regions. Namely, the steric # is four: three bonding pairs and one nonbonding pair.

3. VSEPR theory forces the four electron groups to be placed at the corner of **a tetrahedron**. Namely, it generates an electronic geometry of tetrahedron.

4. Since only three terminal H are available for bonding, the molecular geometry becomes a **trigonal pyramid**.

5. This molecule is not symmetrical with respect to a point.

6. Therefore, it is polar.

PROCEDURE

The objective of this exercise is to predict the structure and simple properties of a molecule based on its Lewis electron dot formula and the Valence Shell Electron Pair Repulsion (VSEPR) theory.

For the given molecules, apply the following guidelines and label as shown in Figure 4.1.

1. Draw the Lewis electron dot formula. The least electronegative atom should be placed at the center of the molecule. (The H atom is an exception. It must always be at a terminal position.)

2. Find the steric # (the number of electron-dense regions surrounding the central atom). Identify the number of bonding pairs of electrons and the number of nonbonding pairs of electrons around the central atom.

3. First, apply the VSEPR theory to determine its electronic geometry.

# of e⁻ Sets	Electronic Geometry	Orbital Types	Model
2	Line	sp	
3	Trigonal Planar	sp^2	
4	Tetrahedron	sp^3	4-hole ball
5	Trigonal Bipyramid	sp^3d	5-hole ball
6	Octahedron	sp^3d^2	6-hole ball

4. Select the ball in the model kit that represents the appropriate electronic geometry.

5. Add surrounding atoms to the bonding electron pairs in the central atom to have a particular 3-D structure (molecular geometry). Use the balls in the model kit to attach the surrounding atoms to the electronic sticks accordingly.

CO$_2$		
Central Atom	C	
Steric Number	2	$:\ddot{O}=C=\ddot{O}:$
Electronic Geometry	linear	
# of Bonding Pairs	2	
# of Lone Pairs	0	
Molecular Shape	linear	

FIGURE 4.1 Notice that a double bond counts as one electron-dense region. Therefore, CO$_2$ has a steric number of 2. This accounts for each double bond.

Name _____ Date _____

Instructor _____

REPORT SHEET | EXPERIMENT

Structure of Molecules | 4

PBr$_5$	
Central Atom	
Steric Number	
Electronic Geometry	
# of Bonding Pairs	
# of Lone Pairs	
Molecular Shape	

HCl	
Central Atom	
Steric Number	
Electronic Geometry	
# of Bonding Pairs	
# of Lone Pairs	
Molecular Shape	

NH$_4^+$	
Central Atom	
Steric Number	
Electronic Geometry	
# of Bonding Pairs	
# of Lone Pairs	
Molecular Shape	

XeF$_4$	
Central Atom	
Steric Number	
Electronic Geometry	
# of Bonding Pairs	
# of Lone Pairs	
Molecular Shape	

SnCl$_6^{2-}$	
Central Atom	
Steric Number	
Electronic Geometry	
# of Bonding Pairs	
# of Lone Pairs	
Molecular Shape	

SCl$_4$	
Central Atom	
Steric Number	
Electronic Geometry	
# of Bonding Pairs	
# of Lone Pairs	
Molecular Shape	

H$_3$O$^+$	
Central Atom	
Steric Number	
Electronic Geometry	
# of Bonding Pairs	
# of Lone Pairs	
Molecular Shape	

H₂CO

Central Atom	
Steric Number	
Electronic Geometry	
# of Bonding Pairs	
# of Lone Pairs	
Molecular Shape	

IF₅

Central Atom	
Steric Number	
Electronic Geometry	
# of Bonding Pairs	
# of Lone Pairs	
Molecular Shape	

CCl₄

Central Atom	
Steric Number	
Electronic Geometry	
# of Bonding Pairs	
# of Lone Pairs	
Molecular Shape	

BH₂⁻

Central Atom	
Steric Number	
Electronic Geometry	
# of Bonding Pairs	
# of Lone Pairs	
Molecular Shape	

1. Predict the F—B—F bond angle in BF_3. Draw a structure to illustrate.

2. Predict the F—S—F bond angle in SF_6. Draw a structure to illustrate.

3. Why is the H—O—H bond angle in a water molecule (H_2O) smaller than the H—C—H bond angle in methane (CH_4)?

4. Draw the structure for $SCl_3F_2^+$. Place the F atoms in the axial positions and the Cl atoms in the equatorial positions. Label all bond angles.

A Gravimetric Analysis of Phosphorus in Fertilizer | 5

As consumers, we are bombarded by images and slogans created by marketing associates. Often these descriptions and claims can be misleading. For example, you have probably seen a bottle or can of juice that has a label stating "contains 110% vitamin C." Clearly, the juice is not composed of 110% vitamin C. Vitamin C is ascorbic acid, a water-soluble vitamin with the molecular formula $C_6H_8O_6$. The phrase "contains 110% vitamin C" refers to the fact that the juice contains 110% of the daily recommended allowance of vitamin C required for an adult to maintain a healthy diet.

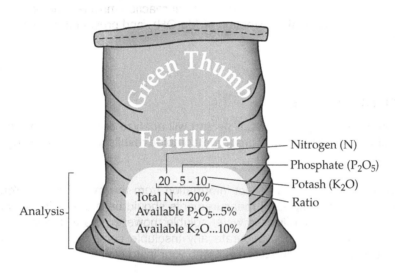

FIGURE 5.1 Conventional fertilizer labeling.
Adapted from Experiments in General Chemistry Laboratory Manual, by Daphne Norton, Hayden McNeil Publishing.

Another example of inaccurate labeling is seen on fertilizer packaging as shown in Figure 5.1. Traditionally, fertilizer contains a rating that represents the nitrogen-phosphorus-potassium content. These elements are important nutrients for plant growth. **Nitrogen** helps plant foliage to grow strong. **Phosphorus** helps flowers and fruit grow and develop. **Potassium** is important for strong root growth and overall plant health.[1] A fertilizer labeled 20-5-10 indicates 20% nitrogen, 5% phosphorus, and 10% potassium by mass. However, this labeling system is misleading because it does not reflect the actual composition of the fertilizer.

The phosphorus and potassium content is expressed as a percentage of P_2O_5, diphosphorus pentoxide, and K_2O, potassium oxide, even though these compounds are not present in fertilizer. Originally, the phosphorus and potassium content was determined by heating a sample of fertilizer in air. At high temperatures, the fertilizer would burn to produce P_2O_5 and K_2O.

Scientists attempt to provide accurate descriptions of matter. Most consumer fertilizers are mixtures of water-soluble salts. Potassium is often present in the form of potassium chloride, while phosphorus is present in the form of a phosphate salt. In the laboratory, you will analyze the phosphorus content of a sample of fertilizer and prepare a new package label to accurately reflect the phosphorus contained in the bag. In order to analyze the phosphorus content of fertilizer, it is necessary to physically separate the phosphorus from all other components in the sample through a physical separation technique called gravimetric analysis.

The phosphorus can be separated from the mixture by a precipitation reaction. The phosphate present in the fertilizer will be converted to hydrogen phosphate and then reacted with Mg^{2+} to form a precipitate, $MgNH_4PO_4 \cdot 6\,H_2O$, as seen in the reaction below.

$$5\,H_2O(l) + HPO_4^{2-}(aq) + NH_4^+(aq) + Mg^{2+}(aq) + OH^-(aq) \rightarrow MgNH_4PO_4 \cdot 6\,H_2O(s)$$

In the laboratory, we can use magnesium sulfate as the source of the Mg^{2+} cation. Magnesium sulfate heptahydrate, $MgSO_4 \cdot 7\,H_2O$, is commercially available and readily soluble in water. The process described above is a reversible reaction. In order to achieve the desired product, $MgNH_4PO_4 \cdot 6\,H_2O(s)$, we must add an excess of magnesium sulfate heptahydrate to create conditions that favor the product side of the reaction.

Also, the reaction must be done under basic conditions for the precipitate to form. The concentration of phosphate ion decreases in an acidic solution due to the formation of hydrogen phosphate.

$$PO_4^{3-}(aq) + H_3O^+ \rightarrow HPO_4^{2-}(aq) + H_2O(l)$$

This means that we must monitor the pH of the reaction to verify that it is done under basic conditions. We will use ammonia to increase the pH of our reaction mixture. The use of a strong base such as NaOH would result in the formation of $Mg(OH)_2$ and prevent the isolation of pure $MgNH_4PO_4 \cdot 6\,H_2O(s)$.

PROCEDURE

Week 1 Gravimetric Analysis

You will work in a group of four people. Each person will analyze a different brand or type of fertilizer. At the end of the experiment, your group will combine results to present findings for four different types of fertilizer.

First, obtain your fertilizer sample and collect information from the manufacturer. You should obtain between 4–6 grams of sample. All measurements should be made using an analytical balance. The phosphorus in the fertilizer is water-soluble; however, the fertilizer may contain some insoluble residue. You must first separate any insoluble components of the fertilizer. Devise a method for this separation and discuss it with your instructor.

Next, you may chemically separate the phosphorus from the rest of the mixture by the reaction described in the background reading. Your group will need to prepare a 1.00 liter solution of 0.400 M magnesium sulfate heptahydrate. This will be your source of Mg^{2+} for the reaction. Figure 5.2 illustrates the steps needed to prepare the volumetric solutions. Refer to your pre-lab questions to calculate the stoichiometric amount of magnesium sulfate heptahydrate solution needed to react with the phosphorus present in your fertilizer sample. As stated in the background reading, we will add an excess of magnesium sulfate heptahydrate to generate the desired product, $MgNH_4PO_4 \cdot 6\,H_2O(s)$. (You should add 50% more magnesium sulfate solution than the volume calculated.)

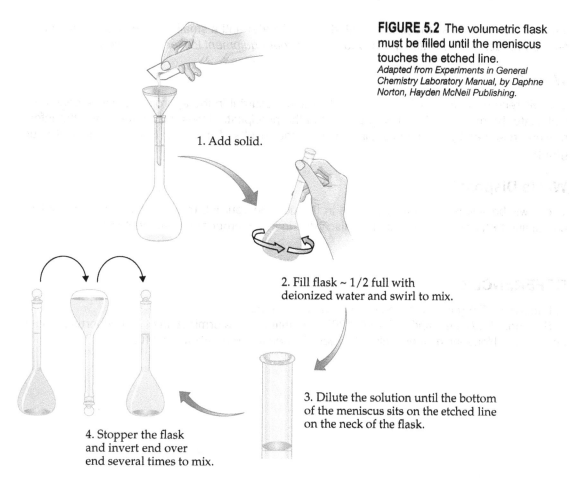

FIGURE 5.2 The volumetric flask must be filled until the meniscus touches the etched line.
Adapted from Experiments in General Chemistry Laboratory Manual, by Daphne Norton, Hayden McNeil Publishing.

1. Add solid.

2. Fill flask ~ 1/2 full with deionized water and swirl to mix.

3. Dilute the solution until the bottom of the meniscus sits on the etched line on the neck of the flask.

4. Stopper the flask and invert end over end several times to mix.

Add the magnesium sulfate heptahydrate solution to your fertilizer mixture, and record your observations. Test the pH of the solution. You will need to add ammonia until the pH of the reaction mixture is approximately 9.0. (*You should add the ammonia slowly and with caution. Ammonia is corrosive and can irritate the skin, eyes, and lungs.*) Recall that you must do the reaction under basic conditions to prevent the formation of hydrogen phosphate. You want PO_4^{3-} available to precipitate out of solution.

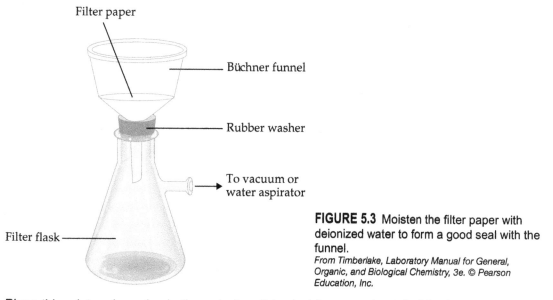

Filter paper

Büchner funnel

Rubber washer

To vacuum or water aspirator

Filter flask

FIGURE 5.3 Moisten the filter paper with deionized water to form a good seal with the funnel.
From Timberlake, Laboratory Manual for General, Organic, and Biological Chemistry, 3e. © Pearson Education, Inc.

Place this mixture in an ice bath, and allow it to cool for approximately 30 minutes. While you are waiting, you can set up a filtration flask that will be used to isolate the precipitate as seen in Figure 5.3. Record the mass of the filter paper and watch glass. Isolate the precipitate using vacuum filtration. Rinse the product with deionized water and a small volume of ethanol. *Transfer the product to your watch glass, and store in your equipment locker until next week.*

Week 2

You will record the mass of your dry product and discard it in the appropriate waste container. Calculate the mass of phosphorus present in the precipitate. Does this value match the information presented by the manufacturer? Collect the results of the other three members of your group.

Waste Disposal

There will be a waste container located in the hood in your lab room. All solutions and solids containing fertilizer residue should be discarded in the appropriate waste container.

REFERENCES

1. http://www.tfi.org/introduction-fertilizer/nutrient-science
2. Solomon, S., A. Lee, and D. Bates. 1993. "Quantitative Determination of Phosphorus in Plant Food Using Household Chemicals." *Journal of Chemical Education* 70: 410.

Name _____ Date _____

Instructor _____

PRE-LAB QUESTIONS | EXPERIMENT

A Gravimetric Analysis of
Phosphorus in Fertilizer | 5

Show all work for partial credit. No credit will be given if no work is shown.

1. A bag of fertilizer is labeled 10-20-20. What is the actual percentage of phosphorus in the fertilizer?

   What is the actual percentage of potassium?

2. Bloom Booster, a fertilizer used for enhancing the blooms of flowering plants, is labeled 10-30-20. Calculate the number of moles of phosphorus, P, in 50.00 grams of this fertilizer.

3. How many grams of magnesium sulfate, $MgSO_4 \cdot 7\ H_2O$, are needed to prepare 250 mL of a 0.400 M solution?

4. What volume of 0.400 M magnesium sulfate is needed to react with HPO_4^{2-} in 15.00 grams of the Bloom Booster fertilizer?

Name ______________________________ Date ______________

Instructor ______________________________

REPORT SHEET | EXPERIMENT

A Gravimetric Analysis of Phosphorus in Fertilizer

5

Student Name	Brand of Fertilizer (manufacturer and type with rating)	% P_2O_5	Calculated % P	Experimental % P

Show all data and calculations for your work:

Mass of fertilizer sample ____________ g

Mass of phosphorus in sample ____________ g

Moles of phosphorus in sample ____________ g

Volume of 0.400 M $MgSO_4 \cdot 7\ H_2O$ ____________ mL

Mass of product   _____ g

Mass of phosphorus in product   _____ g

Percent P in fertilizer   _____ g
(based on isolated product)

The potassium present in fertilizer can react with $MgSO_4 \cdot 7\ H_2O$ to form $MgKPO_4 \cdot 6\ H_2O$ in addition to the desired precipitate. When you calculated the percent mass of P in the fertilizer, you assumed the precipitate was pure $MgNH_4PO_4 \cdot 6\ H_2O$. If some of the solid was actually $MgKPO_4 \cdot 6\ H_2O$, would your calculated percent P be too high, too low, or unaffected? Explain.

Calculate the percent error of the gravimetric analysis of P. (Your calculation will assume that the manufacturer has correctly reported the % P_2O_5.)

$$\frac{\text{Theoretical mass of P in sample} - \text{Experimental mass of P in sample}}{\text{Theoretical mass of P in sample}} \times 100$$

Sustainability Related Questions:

 1. What is the phosphorus cycle?

 2. What critical role does phosphorus play in the structure of plants?

 3. Why do environmentalists worry about the increasing concentration of phosphates in surface waters?

Recycling Aluminum | 6

Chris Niemann/Fotolia

The aluminum beverage can is one of the world's most recycled commercial materials. This is important because the extraction of aluminum from the earth is very energy intensive. Aluminum is the most abundant metal found in the earth's crust. Although it comprises 8.3% of the earth's crust, aluminum is never found in its pure elemental form in nature.[1] It normally exists as bauxite. According to the U.S. Geological Survey, bauxite is a naturally occurring, heterogeneous material composed primarily of one or more aluminum hydroxide minerals, plus various mixtures of silica, iron oxide, titania, aluminosilicate, and other impurities in minor or trace amounts.[2]

Aluminum's durability and corrosion resistance add to its valuable metallic properties of being ductile, malleable, and conductive. The useful physical properties, including its low density, make it worth the effort to extract the metal from the ore. First, aluminum oxide is retrieved from the bauxite when it is washed with a solution of sodium hydroxide at very high temperatures. The aluminum oxide dissolves in the presence of the strong base to produce aluminum hydroxide. The aluminum hydroxide can be recrystallized and later heated to extreme temperatures where it decomposes to form aluminum oxide. The retrieval of aluminum oxide from bauxite is referred to as the Bayer Process.[3]

The aluminum oxide, Al_2O_3, is electrolytically reduced into molten aluminum using the Hall-Heroult process.[3] In this process the aluminum oxide is placed in an electrolytic cell with molten cryolite (sodium aluminum fluoride, Na_3AlF_6). A carbon rod in the cell is charged, and the reaction occurs when the aluminum oxide reacts with the carbon electrode to form carbon dioxide gas and aluminum. The aluminum sinks to the bottom where it is removed from the tank and sent to a melting or holding furnace. The overall reaction for the process is:

$$2\ Al_2O_3 + 3\ C \rightarrow 4\ Al + 3\ CO_2$$

On average, it takes approximately 15.7 kWh of electricity to produce one kilogram of aluminum from bauxite.[4] The recovery of one kilogram of aluminum from a beverage can requires only ~5% of the energy required to produce aluminum from raw materials. This demonstrates how important it is to recycle!

You are probably familiar with the slogan "Reuse, Reduce, Recycle." Today, we will not recycle an aluminum can. Rather, we will reuse it! We will use a beverage can as our source of aluminum. For this experiment, you will utilize consumer resources on campus. You can save a beverage can after you enjoy your favorite drink, or you can collect a can from a local recycling bin. You should rinse the can with hot water before bringing it to lab.

We will save a significant amount of energy by supplying our own starting material rather than ordering it from a chemical manufacturer. We will start with elemental aluminum in its silver, shiny form and convert it to an ion where it will eventually exist as a white powder in the form of $KAl(SO_4)_2 \cdot 12\ H_2O$.

In the first step, aluminum metal will be oxidized to Al^{3+} using a strong base. The result is the formation of an aluminate ion, $Al(OH)_4^-$, and the production of hydrogen gas, as seen in equation 1.

Equation (1) $2\ Al(s) + 2\ KOH(aq) + 6\ H_2O(l) \rightarrow 2\ KAl(OH)_4(aq) + 3\ H_2(g)$

During this reaction, the hydrogen gas escapes into the atmosphere. Now, a strong acid is added, resulting in the formation of aluminum hydroxide, as seen in equation 2.

Equation (2) $2\ KAl(OH)_4(aq) + H_2SO_4(aq) \rightarrow 2\ Al(OH)_3(s) + 2\ H_2O(l) + K_2SO_4(aq)$

Further reaction with acid will cause the $Al(OH)_3$ precipitate to dissolve, leaving Al^{3+} ions in solution, as shown in equation 3.

Equation (3) $2\ Al(OH)_3(s) + 3\ H_2SO_4(aq) \rightarrow Al_2(SO_4)_3(aq) + 3\ H_2O$

The aluminum ions combine with potassium and sulfate ions to crystallize as a hydrated complex. Most of the time, the reaction produces $KAl(SO_4)_2 \cdot 12\ H_2O$.

EXPERIMENTAL PROCEDURE FOR WEEK 1

Your aluminum can should be rinsed with hot water before beginning the experiment. Using scissors, carefully cut a large strip of aluminum from the clean can. *Caution: The metal edges can be very sharp, so be careful not to cut your hand.*

You may use steel wool or sandpaper to remove any paint from the aluminum strip. The aluminum should have a total mass between 1.0 and 1.5 g. (The mass should be recorded to value seen on the analytical balance.) Be sure that you record the mass after removing the paint.

Place the aluminum strip into a 150 mL beaker. Cut the aluminum strip into smaller pieces to increase the surface area and allow it to dissolve more quickly. Now, add ~50 mL of 2.0 M KOH to the beaker containing the aluminum pieces. *Caution: Potassium hydroxide is caustic. The KOH should be added in a fume hood. Use your watch glass to cover the beaker containing the mixture.*

Carefully transfer the beaker to a hot plate. Stir the mixture occasionally with your stirring rod to keep the Al pieces from floating to the top. When it appears that the metal has dissolved, remove the beaker from the hot plate using tongs. Allow the solution to cool, and then remove any solid residue using gravity filtration. Fold your filter paper and set up your funnel, as illustrated in Figure 6.1. Collect the filtrate in your 250 mL beaker. Rinse and dry your watch glass.

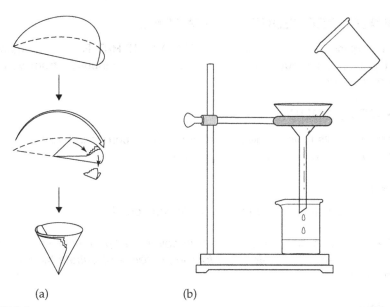

(a)   (b)

FIGURE 6.1 Fold the filter paper and secure a good fit with the funnel before adding solution.
(a) From Tro, Laboratory Manual for Chemistry: A Molecular Approach 3e. © Pearson Education, Inc.
(b) From Timberlake, Laboratory Manual for General, Organic, and Biological Chemistry, 3e. © Pearson Education, Inc.

Slowly add ~30 mL of 6.0 M H_2SO_4 to the filtrate. *Caution: H_2SO_4 is a strong acid. If any acid comes in contact with your skin, notify your instructor immediately and rinse the area with water.*

You should observe a white precipitate. Next, place the beaker on the hot plate and heat gently until all of the white precipitate has dissolved. Remove the beaker from the hot plate using tongs, and allow the solution to cool.

In the meantime, set up an ice bath and place the 250 mL beaker containing your product in the ice bath. Begin setting up a Buchner funnel for vacuum filtration as shown in Figure 6.2.

Pre-weigh a watch glass and filter paper (7.0 cm diameter). Isolate your crystals using vacuum filtration. Transfer your crystals and filter paper to the watch glass, and store in your locker or designated area until next week.

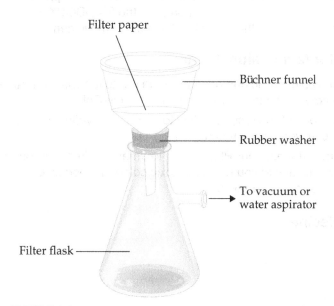

Filter paper

Büchner funnel

Rubber washer

To vacuum or
water aspirator

Filter flask

FIGURE 6.2 Set up the filtering apparatus and attach the side-arm flask to the aspirator or vacuum.
From Timberlake, Laboratory Manual for General, Organic, and Biological Chemistry, 3e. © Pearson Education, Inc.

EXPERIMENTAL PROCEDURE FOR WEEK 2

Obtain the watch glass holding your product, $KAl(SO_4)_2 \cdot 12\ H_2O$. Record the mass of the watch glass, weight paper, and product. Determine the mass contribution from your product, and calculate the percent yield.

ANALYSIS OF PRODUCT

You will perform two tests to verify the content of your new compound, $KAl(SO_4)_2 \cdot 12\ H_2O$. Each test should be done with a fresh sample of product.

Test for sulfate:

Barium ion reacts with sulfate to form an insoluble compound.

Add a small amount of product, just enough to cover the tip of your spatula, to a test tube. Dissolve the solid in 2–3 mL of water. Test for the presence of sulfate by adding 2–3 drops of 1.0 M barium nitrate.

Test for aluminum:

A red dye, aluminon, is used to test for Al^{3+}. The dye is added to the test solution followed by ammonia. A solid, $Al(OH)_3$, should form. This precipitate will have a distinctive pink color if aluminum ions are present.

$$Al^{3+}(aq) + 3\ NH_3(aq) + 3\ H_2O(l) + \text{aluminon} \rightarrow Al(OH)_3 \cdot \text{aluminon}(s) + 3\ NH_4^+(aq)$$

Add a small amount of product, $KAl(SO_4)_2 \cdot 12\ H_2O$, to a test tube. Dissolve the solid in 2–3 mL of water. Add 2 drops of aluminon to the test tube. Next, add 3 M NH_3 dropwise until the solution is basic. A pink or red precipitate confirms the presence of aluminum.

OPTIONAL SERVICE PROJECT

Last week you converted aluminum from a can into $KAl(SO_4)_2 \cdot 12\ H_2O$, commonly known as alum. You probably read that alum is used as a tanning and pickling agent, but you may not have discovered that alum is also an ingredient in Play-Doh™. It impacts the elasticity.

You can actually take a small sample of alum from your product to create handmade Play-Doh.™ Your instructor may have you prepare the Play-Doh™ in class or allow you to take your product home to prepare the Play-Doh™ in your own kitchen.

Instructions for taking alum home:

1. Bring a resealable plastic bag or small container to lab with you. After you weigh your dry product, remove ~2 tablespoons to use in the Play-Doh™.
2. At home, use one of the recipes below to prepare Play-Doh™. You may find other recipes online. You can work with a friend or in a group.
3. Since you may donate your Play-Doh™ to small children who often place things in their mouths, it is crucial that you use a nontoxic coloring. Food coloring or Kool-Aid™ are fine. No inks, paints, dyes, or makeup.

Play-Doh™ Recipes

Ingredients:

1 1/2 cup flour
1 cup salt
1 Tbsp. powdered alum
1 tsp. cooking oil
1 cup boiling water
food coloring

Combine flour, salt, and alum. Add oil and water. Stir the mixture until it is cool, and then knead in the food coloring. When cooled, store dough in an airtight container or a sealed plastic bag.

Ingredients:

1 1/2 cup flour
1/4 cup salt
1 pkg unsweetened Kool-Aid™ (dry powder)
1 cup boiling water
1 Tbsp. alum
1 1/2 Tbsp. vegetable oil

In a bowl, mix flour, salt, and Kool-Aid™. Stir in water and oil. Knead with hands for about 5 minutes. Store in a sealed plastic bag or container for up to 2 months.

REFERENCES

1. Emsley, John. 2001. *Nature's Building Blocks: An A-Z Guide to the Elements*. Oxford: Oxford University Press.

2. United States Geological Survey. http://www.usgs.gov/

3. Earnshaw, A., and N. N. Greenwood. 1997. *Chemistry of the Elements, 2nd edition*. Oxford: Butterworth-Heinemann.

4. "U.S. Energy Requirements for Aluminum Production: Historical Perspective, Theoretical Limits and Current Practices." Prepared for Industrial Technologies Program Energy Efficiency and Renewable Energy, U.S. Department of Energy, February 2007. http://www1.eere.energy.gov/manufacturing/resources/aluminum/pdfs/al_theoretical.pdf

PRE-LAB QUESTIONS | EXPERIMENT
Recycling Aluminum | 6

1. In this week's experiment, you will reuse an aluminum can to produce a new compound, $KAl(SO_4)_2 \cdot 12\ H_2O$. What are the chemical name and common name for this compound? What are some common uses for this compound?

2. What is a hydrated compound?

3. The background reading explains the series of reactions that take place in the formation of $KAl(SO_4)_2 \cdot 12\ H_2O$. Write the net ionic equation for equation 2.

4. If you purchase Coca-Cola in a can, you will see the recycling emblem with a note that says "please recycle." If you purchase the same beverage in a 20-ounce bottle, you will see the same emblem with the letters PETE below the emblem. What does PETE stand for?

 5. The increased consumption of bottled water is generating large volumes of plastic waste. What percentage of aluminum cans gets recycled every year? What percentage of plastic bottles is being recycled? (Give source of data and year reported.)

 6. Is there a health benefit for consuming bottled water instead of tap water? What are the disadvantages to bottled water? Explain and cite your sources.

Name ______________________ Date ______________

Instructor ______________________

REPORT SHEET | EXPERIMENT

Recycling Aluminum | 6

Experimental Data: Synthesis of $KAl(SO_4)_2 \cdot 12\ H_2O$

Mass of Al ____________

Volume of KOH ____________

Volume of H_2SO_4 ____________

Mass of product ____________

Calculations

Moles of Al ____________

Moles of KOH ____________

Moles of H_2SO_4 ____________

Moles of product ($KAl(SO_4)_2 \cdot 12\ H_2O$) ____________

Limiting reagent ____________

Theoretical yield ($KAl(SO_4)_2 \cdot 12\ H_2O$) ____________

Percent yield ($KAl(SO_4)_2 \cdot 12\ H_2O$) ____________

Comment on possible impurities in your product:

Show all work for determining the percent yield.

1. Write an equation explaining how you confirmed that sulfate is present in your product.

2. How many grams of H_2 gas were produced assuming the aluminum metal reacted completely?

3. What is one possible test you could have done to determine if K^+ is present in your product? Write an equation to represent the outcome of this test if K^+ is indeed present.

Qualitative Analysis— The Detection of Anions | 7

You will complete a series of exercises that will develop your observation and reasoning skills. First, you will evaluate a series of known solutions, and then you will use that information to draw conclusions regarding a solution of unknown content.

The solubility principles for cations and anions will prove useful in predicting behavior in solution.

You will begin by studying anions. The specific series you will study include chloride (Cl^-), carbonate (CO_3^{2-}), iodide (I^-), phosphate (PO_4^{3-}), and sulfate (SO_4^{2-}).[1]

> acetate = CH_3COO^-
> chlorate = ClO_3^-
> nitrate = NO_3^-
> silicate = SiO_4^{2-}
> sulfide = S^{2-}

Common Solubility Rules

1. All compounds containing family IA and NH_4^+ cations are soluble.
2. All common acetates, chlorates, and nitrates are soluble.
3. All common chlorides and iodides are soluble except Ag^+, Hg_2^+, and Pb^{2+}. ($PbCl_2$ is soluble in hot water.)
4. All common sulfates are soluble except Ba^{2+}, Ca^{2+}, Hg_2^+, Pb^{2+}, and Sr^{2+}.
5. All common carbonates, phosphates, and silicates are insoluble except family IA and NH_4^+ compounds.
6. All common sulfides are insoluble except family IA and IIA and NH_4^+ compounds.

Begin by considering the behavior of each anion in water.

Chlorides:

Most chloride salts are soluble in aqueous solution with the exception of Ag^+, Pb^{2+}, and Hg_2^{2+}.

Carbonates:

Most carbonates are only slightly soluble in water. However, the carbonate ion reacts with acids to form carbonic acid, which further dissociates into CO_2 and H_2O. The gaseous CO_2 is observed through bubbling.

$$2\,H^+(aq) + CO_3^{2-}(aq) \rightarrow H_2CO_3(aq) \rightarrow H_2O(l) + CO_2(g)$$

Iodides:

Most iodide salts are soluble in aqueous solution with the exception of Ag^+, Pb^{2+}, and Hg_2^{2+}. Also, I^- ions can be oxidized to I_2. (I_2 produces a brown solution in the presence of I^-.) For example, ferric nitrate solutions contain Fe^{3+} ions, which oxidize iodide ions to I_2, as shown below.

$$2\,Fe^{3+} + 2\,I^- \rightarrow 2\,Fe^{2+} + I_2$$

Phosphates:

Most phosphate salts are only slightly soluble.

Sulfates:

Most sulfate salts are soluble in water. Exceptions include Ba^{2+}, Pb^{2+}, Hg_2^{2+}, Ca^{2+}, and Sr^{2+}.

Observing Reactions

You will combine a series of solutions to detect reactivity with anions. It is important to make clear observations and write an equation to explain the reaction that is observed. In some cases, you will observe a simple precipitation reaction where two aqueous solutions are combined and two anions recombine or metathesize to form an insoluble salt, as shown in Figure 7.1.

— Precipitate

FIGURE 7.1 Precipitation may result in the formation of a solid or the formation of a cloudy mixture.
From Tro, Laboratory Experiments for Chemistry: The Central Science 12e. © Pearson Education, Inc.

As you have noted, most of these anions form insoluble salts with Ag^+ ions. It will be crucial to note the color of the salt as this may reveal its identity. Also, some water-insoluble salts will dissolve in an acidic solution. For example, CO_3^{2-} and PO_4^{2-} both react in acid to form soluble components.

$$Ag_3PO_4(s) + 2\,H^+(aq) \rightarrow 3\,Ag^+(aq) + H_3PO_4(aq)$$

Procedure

You will be provided with dropper bottles containing various aqueous solutions. These solutions will be used to develop a chart of expected results. You will also be given a test tube containing an aqueous solution. Your solution may contain one, two, or three of the ions studied in this experiment: Cl^-, CO_3^{2-}, I^-, PO_4^{3-}, and SO_4^{2-}.

Organization and good record keeping are important. Solubility is best evaluated using test tubes, but a spot plate, as illustrated in Figure 7.2, may be useful when analyzing precipitates. When testing for precipitates, use two to three drops of each test reagent. Equal volumes of each reagent should be used unless testing for solubility. How much should be used if testing for solubility?

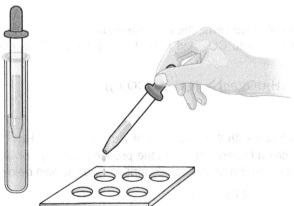

FIGURE 7.2 Carefully label the spot plate, and use equal volumes of test reagents.
Adapted from Experiments in General Chemistry Laboratory Manual, by Daphne Norton, Hayden McNeil Publishing.

	HNO$_3$	AgNO$_3$	AgNO$_3$* + HNO$_3$	Ba(NO$_3$)$_2$	Ba(NO$_3$)$_2$* + HNO$_3$	Fe(NO$_3$)$_3$
(NH$_4$)$_2$CO$_3$						
NaCl	NR			NR	NR	NR
NaI				NR	NR	
Na$_3$PO$_4$	NR					
Na$_2$SO$_4$	NR		NR			NR

*Add two drops of dilute HNO$_3$ to the reaction mixture from columns to the left and mix.
NR = No visible reaction is observed.

REFERENCES

1. Basolo, Fred. 1980. *Journal of Chemical Education* 57 (11): 761.
2. Toby, Sidney. 1995. *Journal of Chemical Education* 72 (11): 1008.

Name ______________________________ Date ______________________

Instructor ______________________________

REPORT SHEET | EXPERIMENT

Qualitative Analysis— The Detection of Anions

7

Anion Unknown Number ________

One to three anions are present in your unknown. Indicate which ion(s) are present by marking the appropriate box.

Ion	Present	Absent
Cl^-		
I^-		
CO_3^{2-}		
PO_4^{3-}		
SO_4^{2-}		

Write *balanced equations* for the following reactions. Use the appropriate subscript to designate the form of the material. (*g*)—gas, (*aq*)—aqueous, (*s*)—solid, etc. Write NR if no reaction occurs.

1. _____ $Ag^+(aq)$ + _____ $CO_3^{2-}(aq)$ →
2. _____ $Ba^{2+}(aq)$ + _____ $Cl^-(aq)$ →
3. _____ $Na_2SO_4(aq)$ + _____ $Ba(NO_3)_2(aq)$ →
4. _____ $AgNO_3(aq)$ + _____ $NaCl(aq)$ →
5. _____ $Fe(NO_3)_3(aq)$ + _____ $Na_2SO_4(aq)$ →
6. _____ $AgNO_3(aq)$ + _____ $NaI(aq)$ →
7. _____ $Ba(NO_3)_2(aq)$ + _____ $NaI(aq)$ →
8. _____ $Ba_3(PO_4)_2(s)$ + _____ $H^+(aq)$ →
9. _____ $HNO_3(aq)$ + _____ $(NH_4)_2CO_3(aq)$ →
10. _____ $HNO_3(aq)$ + _____ $Na_2SO_4(aq)$ →

Qualitative Analysis— Detection of Metal Cations | 8

You will continue to study ions. This experiment will involve the identification of metal cations. You will first analyze the properties of metal cations and then study an unknown solution containing two or three of the following: Ag^+, Al^{3+}, Ca^{2+}, Cu^{2+}, K^+, Ni^{2+}, and Pb^{2+}.

Metal cations demonstrate a variety of characteristics unique to that particular metal. We will utilize these characteristics to distinguish one metal from another. Several metal ions will react when heated in a flame to produce a colored flame. The light of a characteristic color results when the metal ion is heated to high temperatures and absorbs energy to excite electrons to a higher energy state. When these excited electrons relax and drop down back to their original energy state, they emit a light associated with the wavelength of that energy change.

Flame Tests

In order to observe the color emitted from a solution of a metal salt, the solution must be introduced into the hot portion of the Bunsen burner flame, as shown in Figure 8.1. Upon introduction, the colored flame will burn until the metal cation solution is consumed. Liquid samples must be introduced to the flame using a platinum or nichrome wire. This is critical as these types of wire will not interfere with the test. For example, copper ions will generate a blue-green flame. A copper wire would not be appropriate for performing a flame test.

The wire will have a small loop at the end. The wire should be cleaned to verify that it does not contain residue from a previous test. First, place the wire in the flame and observe. You should only observe the natural flame from the Bunsen burner. Next, the loop will be dipped in a concentrated solution containing the metal ion you are attempting to study and placed in the flame. Carefully observe the changes in the flame.

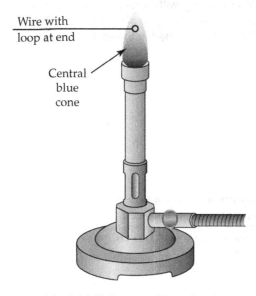

Wire with loop at end

Central blue cone

FIGURE 8.1 Place the nichrome wire in the flame to burn away any residue before testing your unknown sample.
From Timberlake, Laboratory Manual for General, Organic, and Biological Chemistry, 3e. © Pearson Education, Inc.

The most important rule in doing the flame tests is to be aware of the color of emission due to both the clean wire and the known sample. If you cannot distinguish emission of the known solution from that of the clean wire, you cannot use the tests to identify an unknown.

Flame Test Results

Metal Cation	Colored Flame
Ca^{2+}	Red
K^+	Lavender, purple
Cu^{2+}	Blue-green
Na^+	Red
Pb^{2+}	Light blue

Solubility

Separation by precipitation is commonly used in qualitative analysis. It is necessary to know something about the solubility of a variety of salts. Again, you will refer back to your solubility rules.

Typically when trying to separate two kinds of ions in a mixture, it is useful to add a reagent that will cause one ion to precipitate as a salt and allow the other ion(s) to remain in solution. For example, HCl can be added to a mixture of Ag^+ ions and Cu^{2+} ions. The silver ion will precipitate as AgCl, and the Cu^{2+} ion will remain in solution. The precipitate may be separated from the solution by centrifugation.

Heating and Evaporation

It may be necessary to heat a solution to verify solubility or concentrate the solution by evaporating some of the water. It is unsafe to heat the test tubes directly over a flame, as the liquid may squirt out of the tube. Test tubes may be heated in a hot water bath. Your instructor will demonstrate the setup used for holding the test tube in place in the water bath. You may use a clamp or test tube holder to keep the test tube in place.

Using a Centrifuge

Your instructor will demonstrate the use of a centrifuge, the instrument shown in Figure 8.2. The solution being tested will be placed in the centrifuge along with a water blank to balance the centrifuge. *Caution: Do not stick your hand in the centrifuge to try to slow it down. Allow the instrument to come to rest on its own.* Any solid precipitate that formed in the test solution will be

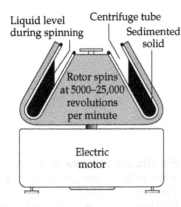

FIGURE 8.2 The centrifuge must be properly balanced.
From Tro, Laboratory Experiments for Chemistry: The Central Science 12e.
© Pearson Education, Inc.

compacted in the bottom of the tube. The solution portion may be decanted or removed using a pipet or dropper. After centrifugation, the solid and liquid components have been isolated. It may be necessary to wash the solid precipitate.

Conditions for Precipitation

When using precipitation as a separation technique, one should consider the following: (1) conditions for precipitation, (2) completeness of precipitation, (3) the problem of coprecipitation, and (4) the separation of solid and liquid phases.

First, precipitation may occur upon proper mixing of two solutions. You should not assume the solutions will mix well when drops of a test reagent are added. It may be necessary to swirl or stir the mixture to achieve good mixing. Failure to adequately mix solutions may result in erroneous conclusions.

The acidity of the solution may also affect the separation of ions. In particular, anions of weak acids such as S^{2-}, CO_3^{2-}, and PO_4^{3-} will frequently precipitate metal ions from neutral or basic solutions but not from an acidic solution. This occurs because the concentration of anions may be very low in the acid solution since the anions are readily converted to weak acids.

Incomplete Precipitation

In order to achieve complete precipitation, the salt that forms must be insoluble in the solution used. The choice of precipitating reagent, concentration, and conditions is important. Sufficient precipitation reagent must be added. Too little may not produce visible precipitation, but too much may cause the precipitate to redissolve or be difficult to observe.

To check for complete precipitation, test the mixture after placing it in the centrifuge. After the centrifuge has been used to separate the solid and liquid, add a drop or two of the test reagent to the clear liquid solution. If the solution becomes cloudy, this indicates that precipitation was incomplete, as the cloudiness represents additional precipitation. Continue this process until you are certain that this particular ion has been completely removed from the solution via precipitation.

Coprecipitation

Coprecipitation occurs when the precipitation of one salt causes the simultaneous precipitation of another salt that would not normally precipitate. For example, Al^{3+} ions react with aqueous ammonia (ammonium hydroxide solution) to form solid $Al(OH)^3$, while calcium ions, Ca^{2+}, are not precipitated. However, when a solution containing both Al^{3+} and Ca^{2+} is treated with aqueous ammonia, the result is a gelatinous $Al(OH)_3$. Most likely the gelatin precipitate contains $Ca(OH)_2$ as well. This will interfere with the analysis of the sample. The best method to correct this situation is by washing the precipitates. After isolating the solid by centrifugation, wash the solid with 5 drops of distilled water, mix well, and centrifuge again. This treatment will dissolve the coprecipitated salt, in this case, $Ca(OH)_2$. Some situations may require two to three washings.

Understanding the Test Reagents

HCl:

In qualitative analysis, an aqueous solution of HCl can be used as a source of H^+ and Cl^- ions. It can produce an acidic environment, or it can react with metal ions to form insoluble chloride salts. Only a few metal ions react to form insoluble salts. Pb^{2+} reacts with 2 Cl^- to form $PbCl_2(s)$.

NH$_3$:

Aqueous ammonia is a source of NH_3 molecules and may also serve as a source of hydroxide ions. ($NH_3 + H_2O \rightarrow NH_4^+ + OH^-$) The presence of OH^- will cause the precipitation of a number of metal hydroxides, such as $Pb(OH)_2$. Since ammonia produces a relatively low concentration of OH^- ions, it will not precipitate moderately soluble $Ca(OH)_2$ and does not dissolve $Pb(OH)_2$, which will normally dissolve in strong bases such as NaOH.

Some metal hydroxides such as $Cu(OH)_2$ and some insoluble salts such as AgCl will dissolve in an ammonia solution by forming complex ions with the NH_3 molecules. $[Cu(NH_3)_4]^{2+}$ and $[Ag(NH_3)_2]^+$ are examples.

NaI:

Sodium iodide serves as a source of I^- ions that in some cases will form insoluble salts with metal cations as with AgI. The iodide ion is also a mild reducing agent and can be oxidized to a red-brown I_3^- ion by oxidizing agents such as Cu^{2+}.

NaOH:

A sodium hydroxide solution provides a high concentration of hydroxide ions. This can be used to precipitate numerous metal hydroxide compounds. However, as a result of the high concentration of OH^- ions that are produced, an excess of NaOH will dissolve some hydroxides such as $Pb(OH)_2$ and form soluble hydroxide complexes as with $[Pb(OH)_3]^-$.

$(NH_4)_2(CO)_3$:

Many metal ions will form insoluble carbonate salts when treated with an aqueous solution of $(NH_4)_2(CO)_3$. Carbonate solutions also contain hydroxide ions.

$$CO_3^{2-}(aq) + H_2O(l) \rightarrow HCO_3^-(aq) + OH^-(aq)$$

A hydroxide compound may be precipitated in preference to a carbonate, as is the case with $Al(OH)_3$. Hydrolysis of the NH_4^+ ion generates NH_3, which forms soluble complexes such as $Cu(NH_3)_4^{2+}$ and $Ag(NH_3)_2^+$.

Properties of Metal Ions

Ag^+, silver ion:

- Forms insoluble salts with chloride, iodide, sulfide, and carbonate ions.
- Silver oxide is precipitated by NaOH and does not dissolve in excess NaOH.
- Silver ions form a stable ammine complex, $[Ag(NH_3)_2]^+$.

Al^{3+}, aluminum ion:

- Forms an insoluble hydroxide that is amphoteric.
- Forms $Al(OH)_3$ when treated with a base. The product is a gelatinous solid.
- $Al(OH)_3$ readily dissolves in a solution containing a high OH^- concentration.

Ca^{2+}, calcium ion:

- Forms insoluble carbonate salt. Calcium carbonate is soluble in acidic solution.
- Forms calcium hydroxide that is slightly soluble. Calcium hydroxide will not be precipitated in the presence of a weak base, ammonia.
- Gives a red flame test, but color may be difficult to detect.

Cu^{2+}, copper (II) ion:

- Gives a blue-green flame test.
- Forms insoluble sulfide and carbonate salts.
- Precipitates copper hydroxide in the presence of NaOH.
- Copper hydroxide dissolves in an ammonia solution by forming $[Cu(NH_3)_4]^{2+}$.
- Reacts with excess iodide ions to form insoluble CuI. Copper is reduced to Cu(I) while the iodide ion is oxidized to red-brown triiodide ions. (The color of the I_3^- ions can be removed by heating the solution in a beaker to drive off I_2 from I_3^- leaving I^-. Also, the color can be removed by adding base that converts I_3^- to I^- and IO_3^-, both of which are colorless.)

K^+, potassium ion:

- Potassium ions form no insoluble salts.
- Gives a purple color in the flame test.

Ni^{2+}, nickel (II) ion:

- Hydrated nickel ion, $[Ni(H_2O)_6]^{2+}$ is a pale green color.
- Forms an insoluble hydroxide, $Ni(OH)_2$. Nickel (II) hydroxide does not readily dissolve in an NH_3 solution.
- Forms a stable complex ion with ammonia, $[Ni(NH_3)_6]^{2+}$.
- Forms an insoluble carbonate compound.
- Forms an insoluble sulfide compound. The sulfide will not easily dissolve in acid and will not precipitate from a solution that is strongly acidic.
- To clarify between copper and nickel ions, dimethylglyoxime can be used. In ammonia, nickel forms as nice pink precipitate with one drop of a solution of dimethylglyoxime. Large concentrations of copper ion may interfere with this test. Then how does it differentiate between them?

Pb^{2+}, lead (II) ion:

- Forms insoluble chloride, iodide, sulfide, and carbonate salts.
- Lead chloride forms slowly and is somewhat soluble in water.
- Lead hydroxide is insoluble but amphoteric. It is precipitated by ammonia but readily dissolves in NaOH to form $[Pb(OH)_3]^-$.
- Gives a light blue color in the flame test.

PROCEDURE

Identification Guidelines

The following guidelines will assist you in achieving accurate results.

1. Clearly label all test tubes and precipitates and record observations immediately. Do not rely on memory.
2. Read the labels of test reagents carefully.
3. Do not contaminate test solutions by putting pipets or droppers into the bottles. If a test reagent becomes contaminated, it must be discarded.
4. Use clean glassware.
5. After adding a test reagent, stir to ensure proper mixing.
6. The acidity of solutions is important. Check the acidity with pH paper when necessary.
7. Test for completeness of precipitation.
8. Wash precipitates to avoid coprecipitation.
9. It is always useful to refresh your memory of the solubility rules.

Common Solubility Rules

1. All compounds containing family IA and NH_4^+ ions are soluble.
2. All common acetates, chlorates, and nitrates are soluble.
3. All common chlorides are soluble except Ag^+, Hg^{2+}, and Pb^{2+}. ($PbCl_2$ is soluble in hot water.)
4. All common sulfates are soluble except Ba^{2+}, Ca^{2+}, Hg^{2+}, Pb^{2+}, and Sr^{2+}.
5. All common carbonates, phosphates, and silicates are insoluble except family IA and NH_4^+ compounds.
6. All common sulfides are insoluble except family IA and IIA and NH_4^+ compounds.

	NH$_3$*	(NH$_4$)$_2$CO$_3$*	NaOH*	NaI	HCl	flame test
AgNO$_3$						
Al(NO$_3$)$_3$						
Ca(NO$_3$)$_2$						
Cu(NO$_3$)$_2$						
Ni(NO$_3$)$_2$						
Pb(NO$_3$)$_2$						
KNO$_3$	NR	NR	NR	NR	NR	NR

* Test the results of any precipitates that form.
Specifically, test precipitates formed with NH$_3$ for solubility in NaOH.
Test precipitates formed by reaction with (NH$_4$)$_2$CO$_3$ for solubility in HNO$_3$.
Test precipitates formed by reaction with NaOH for solubility in NH$_3$.
NR = No visible reaction is observed.

REFERENCES

1. Basolo, Fred. 1980. *Journal of Chemical Education* 57 (11): 761.
2. Toby, Sidney. 1995. *Journal of Chemical Education* 72 (11): 1008.

Name _____ Date _____

Instructor _____

REPORT SHEET | EXPERIMENT

Qualitative Analysis—Detection of Metal Cations | 8

Metal Cation Unknown Number _____

Two to three cations are present in your unknown. Indicate which ions are present by marking the appropriate box (*8 points each*):

Ion	Present	Absent
Ag^+		
Al^{3+}		
Ca^{2+}		
Cu^{2+}		
Ni^{2+}		
Pb^{2+}		
K^+		

1. If a solution contains Cu^{2+} and Ni^{2+} ions, what test reagent can be used to separate these ions? *Write equations indicating the results.*

2. Write the net ionic equation of $AgNO_3(aq)$ reacting with $(NH_4)_2CO_3(aq)$.

3. If a solution contains Ag^+ and Ca^{2+} ions, what *test reagent* can be used to separate these ions? *Write equations indicating the results.*

4. Write the net ionic equation for $Ni(NO_3)_2$ reacting with NaOH. Give the formula and name of any precipitates that form.

5. Explain why NaOH is not the best test reagent for detecting Pb^{2+}.

6. Write the net ionic equation for $Ca(NO_3)_2(aq)$ reacting with NaOH(aq).

Qualitative Analysis— Identification of a Single Salt | 9

To continue our qualitative analysis, you will be given a metal salt and will be required to identify both the cation and anion. Experience from the previous two lab exercises will be valuable.

Common Solubility Rules

1. All compounds containing family IA and NH_4^+ cations are soluble.
2. All common acetates, chlorates, and nitrates are soluble.
3. All common chlorides are soluble except Ag^+, Hg_2^+, and Pb^{2+}. ($PbCl_2$ is soluble in hot water.)
4. All common sulfates are soluble except Ba^{2+}, Ca^{2+}, Hg_2^+, Pb^{2+}, and Sr^{2+}.
5. All common carbonates, phosphates, and silicates are insoluble except family IA and NH_4^+ compounds.
6. All common sulfides are insoluble except family IA and IIA and NH_4^+ compounds.

PROCEDURE: EVALUATING SOLUBILITY

Begin by testing the solubility properties of your unknown. Use a small portion of your unknown to determine if your salt is soluble in water.

If your salt is insoluble in water, try heating the solution by placing your test tube in a hot water bath.

If your salt is insoluble in water even upon heating, try dissolving the salt in 2M HNO_3.

If the salt is insoluble in water but dissolves in acid, the anion is probably derived from a weak acid. Thus, the anion is *phosphate* or *carbonate*. Some *sulfates* also readily dissolve in acid.

If the anion is a *carbonate*, it will be destroyed in the acidic solution. This means you must test for it as the salt dissolves. To observe the CO_2 bubbles that form when *carbonate* reacts with acid, you may mix a sample of salt with ten drops of water and then centrifuge the wet salt to the bottom of the test tube. Next, place five drops of acid on the surface of the solution. You should be able to see bubbles as the acid diffuses down to the bottom of the tube and reaches the salt.

Phosphate salts are also insoluble in water. They dissolve in acid but may re-precipitate when neutralized. To test for *phosphate* in an acidic solution, add four drops of ammonium molybdate solution $(NH_4)_2MoO_4$. Mix the reagents thoroughly, and heat in a hot water bath. The presence of *phosphate* is confirmed by the precipitation of a yellow solid, $(NH_4)_3PO_4 \cdot 12\ MoO_3$. It is crucial to test a standard solution containing phosphate alongside the unknown to confirm the expected precipitate.

This week's exercise involves investigation. It is important to arrive prepared and organized. Proper note taking and careful observation are critical. Avoid cross contamination of samples by using clean glassware and labeled test tubes. Always use distilled or deionized water.

Name _____ Date _____

Instructor _____

REPORT SHEET | EXPERIMENT

Qualitative Analysis—
Identification of a Single Salt | 9

Single Salt Unknown Number _____

One anion and one cation are present in your unknown. Name the anion and cation. (25 points each)

Metal cation _____

Anion _____

Write the net ionic equation for each reaction. If no precipitate forms, you must write NR.

Write the net ionic equation of your metal cation reacting with chloride ion.

Write the net ionic equation of your metal cation reacting with hydroxide ion.

Write the net ionic equation of your anion reacting with silver.

Write the net ionic equation of your metal cation reacting with nitrate.

Write the net ionic equation of your metal cation reacting with carbonate ion.

Qualitative Analysis of Household Chemicals | 10

Daphne Norton

The last three experiments focused on the identification of cations and anions within a sample. Using the same methods for analysis, you will apply those same skills to a practical situation.

Your friend John invites you to visit his family's cabin in northern Maine. You are excited to spend the summer in Maine, but you are surprised by the condition of the cabin. No one has visited the cabin since the 1980s. The place is a wreck! You and John have a lot of cleaning to do. You start in the kitchen where you find a cardboard box full of baby food jars. John's grandmother used the small jars to store cooking supplies and some medications. One jar has small white tablets. The label on the jar says aspirin. Some jars contain spices. Unfortunately, the labels on some of the other jars have fallen off.

You find six jars with no labels, but digging through the box you find seven paper labels that read salt, antacid, baking soda, washing soda, sugar, cornstarch, and Epsom salt.

PRE-LAB QUESTIONS | EXPERIMENT

Qualitative Analysis of Household Chemicals | 10

Devise a method for indentifying the six common household chemicals. All six jars contain white powders.

1. Begin by learning the chemical name, chemical formula, and active ingredient in each of the household chemicals.

Household Chemical	Chemical Name	Formula
salt		
antacid		
baking soda		
washing soda		
sugar		
cornstarch		
Epsom salt		

2. Some of the household chemicals are not ionic compounds. What does this mean?

3. Use the library, your textbook, or the Internet to determine a method for analyzing the compounds that will not dissociate into ions when added to water. (You may need a specific test for each compound.)

4. Complete the chart to recognize the differences and similarities in reactivity.

Household Chemical	Properties/Reactivity
salt	
antacid	
baking soda	
washing soda	
sugar	
cornstarch	
Epsom salt	

5. Devise a detailed procedure or flow chart for analyzing the six white powders.

Name _____ Date _____

Instructor _____

REPORT SHEET | EXPERIMENT

Qualitative Analysis of Household Chemicals | 10

Sample # _____

Sample	Identity	Comments/Observations
A		
B		
C		
D		
E		
F		

Iron Deficiency Analysis | 11

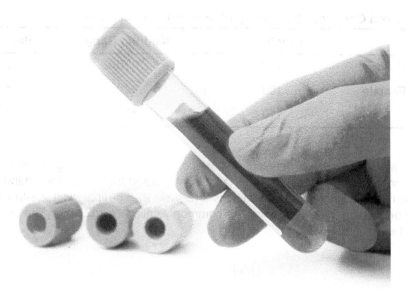

Alexander Raths/Fotolia

Introduction and Background

A very important field of science is pharmaceutical chemistry. Chemists are responsible for the development of medications used to treat and, in some cases, prevent human disease. Typically, the new medications are synthesized by organic and medicinal chemists. Another important aspect of medicine is the screening and detection of infection. Biomolecular chemists develop assays for the detection of pathogens, while analytical chemists devise calibration methods for quantifying the appropriate levels of minerals and enzymes in our blood. Without the skills of these talented chemists, physicians would face even greater challenges when trying to diagnose and treat patients.

Metals in Biology

Iron is an essential component of the human body. Iron is incorporated into heme proteins such as hemoglobin, which is used for oxygen transport. Iron also enables cells to release energy, and it scavenges dangerous free radicals.

Food supplements contain iron in the 2+ oxidation state, but once in the body, the iron oxidizes to the Fe^{3+} state. Conversely, iron can be very toxic, so the ability to store and release iron in a controlled fashion is essential. Cells have solved this problem of iron storage by developing ferritins, a family of iron-storage proteins that sequester iron inside a protein coat. Fe^{3+} ions attached to ferritin are released to transferrin, another protein system present in blood plasma. Transferrin then transports the iron ions to blood-forming sites.[1]

When a person does not have enough iron to meet the body's demand, the individual may suffer from anemia. Anemia is a decrease in the amount of red cells in the blood caused by having too little iron. It is often caused by blood loss or a diet insufficient in iron.

In this week's experiment, you will analyze a patient's blood to determine if the patient has the appropriate level of iron in his/her blood. The level of iron necessary to function is dependent on gender and age. Below you will see a table of iron concentrations found in men, women, and children.[2,3]

Normal Iron Concentrations in μg/dL (micrograms per deciliter)

	Low Range	Medium Range	High Range
Adult Male	120	170	220
Adult Female	100	140	190
Child	110	150	200

Analytical chemists have developed a method for blood screening. They have developed an assay for total serum iron. It is important to understand the chemistry behind the routine method. The reagent used in the screening is 3-(2-pyridiyl)-5,6-diphenyl-1,2,4-triazine-p,p'-disulfonic acid, monosodium salt. The reagent is commonly called FerroZine™. The chemical structure is shown here:

FerroZine™ is a water-soluble salt that forms a 2⁻ anion in solution. The colorless FerroZine™ solution reacts with Fe^{2+} to form a magenta-colored complex. The intensity of the color is directly related to the concentration of Fe^{2+}.[4,5]

$$[Fe(H_2O)_6]^{2+}(aq) + 3\ FerroZine^{2-}(aq) \rightarrow Fe(FerroZine)_3^{4-}(aq) + 6\ H_2O(l)$$

Since FerroZine™ binds only to Fe^{2+} ions, an extra step must be taken to convert any Fe^{3+} ions in solution to Fe^{2+}. Hydroxylamine is used to reduce the iron, as shown in the reaction below.

$$2\ Fe^{3+} + 2\ NH_2OH \rightarrow 2\ Fe^{2+} + N_2 + 2\ H^+ + 2\ H_2O$$

FerroZine™ is used for many applications for the detection of iron. Toxic waste analysis, soil analysis, and water analysis may be done. FerroZine™ even has forensic applications, as trace amounts of blood can be detected when a suspect's hands are sprayed with a solution of FerroZine™.[6]

SAMPLE ANALYSIS: USE OF A SPECTROPHOTOMETER

After the sample has been treated with FerroZine™, we will use a spectrophotometer to measure the absorption of the magenta solution. The basic law that is applicable in this experiment can be stated as follows: *The absorption of light as it passes through a solution is proportional to the concentration of the absorbing species, the length of the light path, and a fundamental property of the material called the molar absorptivity.* See Figure 11.1.

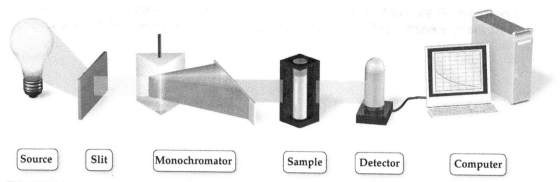

FIGURE 11.1 The mechanics of a spectrometer are shown.
From Tro, Laboratory Manual for Chemistry: A Molecular Approach 3e. © Pearson Education, Inc.

The equation $A = \varepsilon l c$ is a conclusion of the **Lambert–Beer law**. The Greek letter ε represents the molar absorptivity, the letter l the light path length, and c the concentration. A solution that absorbs light in the visible range is colored. The color observed is dependent on the wavelengths of light that pass through the solution. This experiment measures the amount of light absorbed and relates it to the concentration of the absorbing substance.

To better understand the principles of this experiment, some definitions are needed. In the following discussion, the symbol I represents the intensity of light that passes through an absorbing solution and I_o represents the intensity of the incident light. The ratio I/I_o is defined as a quantity called the **transmittance** and is represented by the symbol T. The value T multiplied by 100 is the percent transmittance, **%T**. The device used in the laboratory to select the wavelength and measure the intensity of light passing through a sample is called a *spectrophotometer*. The spectrophotometers used in the chemistry laboratory display **%T** and a second related value, **absorbance, A**. The absorbance is the inverse logarithm of the transmittance. This relationship can be written two ways:

$$A = log(1/T) \text{ or } A = -log(I/I_o).$$

The spectrophotometers in our laboratory have a sensitive meter that is marked in both %T and A. The A scale is logarithmic and is difficult to read accurately at higher values of A, so it is common practice to read the %T scale estimating to the nearest tenth %. The values in %T can easily be converted to A by using the relationship $A = 2 - log\ \%T$. Some digital spectrophotometers display an absorbance reading without requiring a conversion. Your instructor will provide specific instructions and demonstrate use of the spectrometer.

CLINICAL PROTOCOL

Special note: You will not be working with actual blood samples. Instead, you will analyze solutions of iron(II) ammonium sulfate, $Fe(NH_4)_2(SO_4)_2$.

Caution: This lab will require the use of hydroxylamine. Hydroxylamine is toxic and a potential mutagen. Although the solution you will be using is weakly concentrated, you should wear gloves and wash your hands immediately after using this reagent.

Your instructor will demonstrate the proper use of a spectrophotometer. Be careful not to scratch the cuvettes. Turn on the spectrophotometer and let it warm up.

In order to analyze a patient sample, you must first devise a set of standards. You will use the same protocol to analyze the standards and the patient sample. The standards will include blood samples containing the "normal" range of iron concentration. You will also need to prepare one standard above the "normal" range and one standard "below" the normal range.

To reduce the Fe^{3+} ions to the Fe^{2+} form that will react with FerroZine™, you must treat the sample with hydroxylamine. Next, you will add FerroZine™. The later reaction should produce a bright pink color. The intensity of the color is dependent on the concentration of Fe^{2+} present. Please note the concentration of your original sample will change as a result of the new volume.

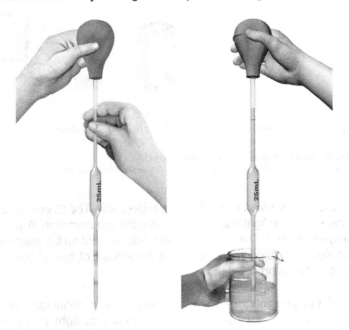

Figure A: Bulb sits lightly on top of pipet. Create suction by squeezing the bulb.

Figure B: Pipet tip must be below surface of the liquid. Fill pipet above the calibration mark.

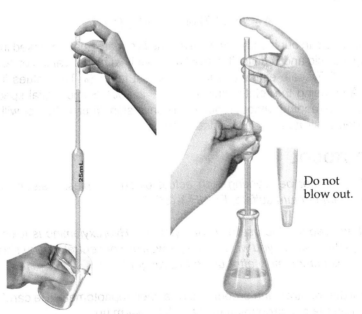

Figure C: Remove pipet bulb and immediately cover end with your index finger. Drain level until bottom of meniscus sits on calibration mark.

Figure D: Transfer liquid. The pipet is calibrated to retain a small amount of liquid.

Do not blow out.

FIGURE 11.2 A volumetric pipet is calibrated to deliver an exact volume of liquid.
From Tro, Laboratory Manual for Chemistry: A Molecular Approach 3e. © Pearson Education, Inc.

Practice pipetting with distilled water before pipetting standards. Refer to Figure 11.2 for the proper technique.

Use a volumetric pipet to dispense exactly 10.00 mL of a standard sample into a small beaker. Next, use a pipet to add 5.00 mL of hydroxylamine stock solution. Mix the contents by swirling. Next, add 5.00 mL of the FerroZine™ stock solution. Again, swirl the contents.

Now, you will use the spectrophotometer to measure the % Transmittance of your sample. (Your instrument may record Absorbance.) Use the dial to select a wavelength of 562 nm. This wavelength gives the maximum absorbance for our sample solutions. See Figure 11.3.

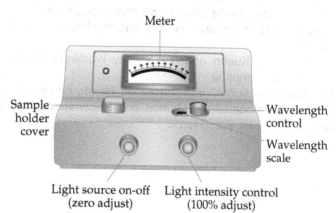

FIGURE 11.3 Use the proper dial to adjust the wavelength
From Tro, Laboratory Manual for Chemistry: A Molecular Approach 3e. © Pearson Education, Inc.

Transfer a small portion of your solution to a cuvette. A cuvette is a special test tube made of quartz. Quartz does not absorb any light, so it will not interfere with your results. The cuvette should be wiped clean before inserting into the spectrometer. The line of the cuvette should match the notch in the spectrophotometer. You will use water as your reference. The pathlength for the cuvettes you will use is 1.00 cm. Record the %T. Record your findings in a table in your notebook.

Repeat the procedure for all standard solutions and your sample.

Waste Disposal

After you have completed your measurements with the spectrophotometer, place the solutions in the waste container that is provided for the diluted solutions.

Data Analysis

Convert your %T readings to Absorbance values. Use Excel or another graphing program to plot the absorbance versus concentration for the standard solutions. (Absorbance is plotted as the *y*-axis and the concentration on the *x*-axis.) This graph is referred to as a Lambert–Beer law plot.

Add a trendline to determine the slope. The slope of the graph is the molar absorptivity, ε, from the Lambert–Beer equation.

Use this information to determine if your patient's iron level is too low or too high.

Treatment of the Patient

After diagnosing your patient as having iron levels that are too low, normal, or too high, you will be asked to make recommendations for your patient's recovery.

If your patient's iron level is in the normal range, you will not need to advise the patient.

Iron(II) sulfate is used in the treatment of iron deficiency anemia. If you need to remove iron in a clinical setting, you would administer an agent that would bind to the iron and cause it to be excreted. Deferoxamine is used in the clinical treatment of acute iron overdose. Deferoxamine has a molar mass of 656.8 g/mol and binds to Fe^{3+} in a one-to-one ratio.

REFERENCES

1. Bertini, Ivano, Harry B. Gray, Stephen J. Lippard, and Joan Selerstone Valentine. 1994. *Bioinorganic Chemistry*. University Science Books.
2. http://www.umm.edu/blood/aneiron.htm
3. http://labtestsonline.org/understanding/analytes/serum-iron/tab/test
4. Stookey, Lawrence L. 1970. Ferrozine—a new spectrophotometric reagent for iron. *Anal. Chem.* 42 (7): 779–781.
5. Wink, D. J., S. Gislason, J. E. Kuehn, and W. D. Hnaey. *Working with Chemistry: A Laboratory Inquiry Program*. Freeman & Company.
6. Lee, C. W. 1986. "The Detection of Iron Traces on Hands by Ferrozine Sprays: A Report on the Sensitivity and Interference of the Method and Recommended Procedure in Forensic Science Investigation." *J. Forensic Sci.*, 31, 11.

Name _____ Date _____

Instructor _____

PRE-LAB QUESTIONS | EXPERIMENT
Iron Deficiency Analysis | **11**

1. When using FerroZine™ to determine the concentration of iron, how many moles of FerroZine™ are needed to complex 3.6×10^{-7} mol of iron?

2. How many mL of FerroZine™ stock solution (100 μg/dL) are needed to obtain the number of moles of FerroZine™ needed in Question 1?

3. How many moles of Fe^{3+} can be reduced by 0.50 mL of 0.434 M NH_2OH?

4. In order to treat a patient, you may have to remove iron from a solution. If you have a 25 mL solution that contains 43 µg/dL of iron, how many grams of iron must be removed to reach a concentration of 23 µg/dL? How many moles of iron is this?

5. In order to treat a patient, you may have to add iron to a solution. If you have 25 mL of a solution that contains 15 µg/dL, how many grams of iron must be added to reach a concentration of 23 µg/dL? How many moles of iron is this?

Name _____ Date _____

Instructor _____

Iron Deficiency Analysis | 11

Patient Sample ID: _____

Part I: Preparing a Calibration Curve

Data Table:

Initial Concentration	Final Concentration	% Transmittance	Absorbance
50 µg/dL	_____	_____	_____
100 µg/dL	_____	_____	_____
110 µg/dL	_____	_____	_____
120 µg/dL	_____	_____	_____
150 µg/dL	_____	_____	_____
170 µg/dL	_____	_____	_____
190 µg/dL	_____	_____	_____
200 µg/dL	_____	_____	_____
220 µg/dL	_____	_____	_____
500 µg/dL	_____	_____	_____

Show the conversion between %T and Absorbance for one of the standards. Show all work and give the equation.

Part II: Graphical Analysis

Attach a graph of absorbance versus concentration. Indicate the molar absorptivity.

Use the Lambert–Beer's law equation to calculate the concentration of iron in your patient's sample. Show all steps of your calculation.

Based on the calculations from your experimental results, is the iron level in your patient's blood too low or too high?

Explain in detail how you would treat your patient's condition. As an example, give the *specific* treatment for a 20.0 mL sample of blood. ("Increase iron level" or "decrease iron concentration" are not sufficient answers.)

Show all necessary calculations.

Gasimetric Analysis of a Carbonate | 12

Gases are a common product of chemical reactions, yet they are difficult to quantify because gases often escape into the atmosphere. Several methods exist for trapping and measuring gases. In this experiment, we will utilize a eudiometer to measure gas volume by water displacement.

Carbonates readily react with acid to produce carbon dioxide. We will perform the following reaction under conditions that allow us to collect the CO_2 and measure the volume of the gas. Knowing the volume, we can apply the ideal gas law to determine n, the number of moles of CO_2 produced.

$$CaCO_3(s) + 2\ HCl(aq) \rightarrow CaCl_2(s) + H_2O(l) + CO_2(g)$$

Dalton's law reminds us that the total pressure of a gas sample is the sum of the partial pressures of all the individual gas components present. In our experiment, CO_2 gas will bubble through a tube containing water. As a result, the gas becomes "wet," meaning the total gas pressure depends on the partial pressure of CO_2 and the water vapor pressure.

We must also consider the height of the water remaining in the eudiometer column. To compare our system to barometric pressure, we must convert the height of the water remaining in the column, Δh, to the equivalent height of a mercury column. A barometer contains a column of mercury. *Hint*: Look up the density of mercury, and the conversion will make more sense. The following equation represents the partial pressure from the CO_2 gas. The barometric pressure is the pressure of the atmosphere.

$$P_{TOT} = P_{atm}$$

$$P_{TOT} = P_{carbon\ dioxide\ gas} + P_{water\ vapor} + \Delta h/13.6$$

$$P_{atm} = P_{carbon\ dioxide\ gas} + P_{water\ vapor} + \Delta h/13.6$$

$$P_{carbon\ dioxide\ gas} = P_{atm} - P_{water\ vapor} - \Delta h/13.6$$

Once the partial pressure of carbon dioxide has been calculated, we can apply the ideal gas law to solve for n. Use the temperature of the room as the value for T.

$$PV = nRT$$

Procedure

As the experiment begins, the instructor will read the barometer and report the atmospheric pressure in the lab room or provide instructions for reading the barometer.

Set up the gas collection apparatus by filling the eudiometer with water and inverting it in a large beaker also filled with water. Cover the end of the eudiometer with your thumb, and submerge it below the water in the beaker. The final setup is shown in Figure 12.1. Before performing the experiment, check to make sure that you do not have air bubbles in the tube. Place the glass tube connected to the rubber tubing inside the eudiometer so that the gas evolved will be transported through the tube into the eudiometer.

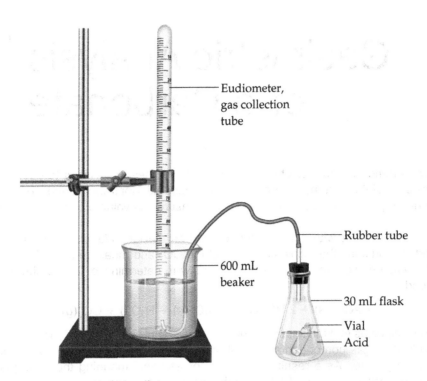

FIGURE 12.1 Check to make sure that the eudiometer is level and that there are no air bubbles in your system.

Measure 0.1–0.15 gram of calcium carbonate in a small, dry vial. Record the exact mass of the solid.

Place ~10 mL of 3 M HCl into the small Erlenmeyer flask. *Caution: Avoid contact with HCl. If you spill acid on your skin, flush with a generous amount of water and immediately contact your instructor.*

Gently place the vial containing calcium carbonate into the flask. Slide the vial down the side of the flask. Your instructor may ask you to wear gloves. Be careful not to mix the solid and the acid. Place the stopper end of the tubing in the Erlenmeyer flask. You have now created a system that will allow the gas from the reaction to travel to the eudiometer without escaping into the atmosphere.

Record the temperature of the water to the tenths place, and also record the temperature in the room. Refer to Table 12.1 presented at the end of the experiment to determine the vapor pressure of water for the temperature you measured for the room.

As shown in Figure 12.2, carefully tip the flask so that a small amount of acid is transferred into the vial. Carbon dioxide should evolve and travel through the tube to the eudiometer. The gas will displace the water in the eudiometer. Gently swirl the reaction vessel until no more gas is evolved. All of the white solid should be consumed. Patiently swirl the contents of the flask for an additional 4–6 minutes after it appears that the reaction has gone to completion. Additional gas may be trapped in solution and will take time to be released.

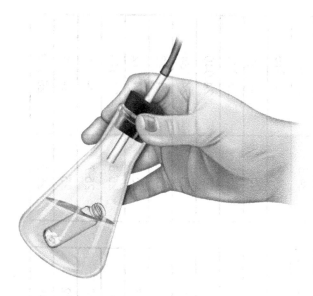

FIGURE 12.2 Check that you have created a closed system before you begin the reaction. Slowly tilt the vial to allow the acid to mix with the solid.
Adapted from Experiments in General Chemistry Laboratory Manual, by Daphne Norton, Hayden McNeil Publishing.

Once the reaction is complete, several measurements must be made. The volume of gas collected may be read to the hundredth place from the eudiometer tube. The height of the water, Δh, should be measured with a ruler as shown in Figure 12.3. The distance from the surface of the water in the beaker to the height of the water level in the eudiometer should be recorded as Δh. The measurement is made in cm but must be converted to mm.

You may discard the contents of the reaction flask into the sink. Rinse with running water. *Caution should be taken since the flask contained 3 M HCl.* Calculate your results for trial 1, and then complete trial 2. You must use a dry vial for each trial.

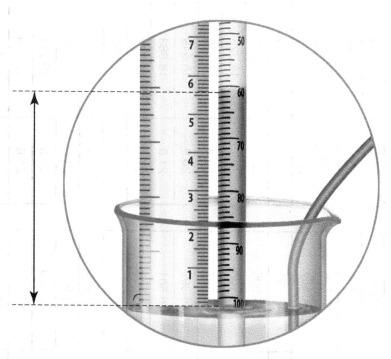

FIGURE 12.3 Use a ruler to measure h, the height of the water.
Adapted from Experiments in General Chemistry Laboratory Manual, by Daphne Norton, Hayden McNeil Publishing.

Table 12.1 Table Vapor Pressure of Liquid Water between 18.0 °C and 25.9 °C

Temp (°Celsius)	Vapor Pressure (torr)	Temp (°Celsius)	Vapor Pressure (torr)	Temp (°Celsius)	Vapor Pressure (torr)	Temp (°Celsius)	Vapor Pressure (torr)
18.0	15.505	20.0	17.552	22.0	19.835	24.0	22.379
18.1	15.603	20.1	17.660	22.1	19.956	24.1	22.513
18.2	15.700	20.2	17.769	22.2	20.078	24.2	22.648
18.3	15.798	20.3	17.879	22.3	20.200	24.3	22.784
18.4	15.897	20.4	17.989	22.4	20.322	24.4	22.921
18.5	15.996	20.5	18.100	22.5	20.446	24.5	23.058
18.6	16.096	20.6	18.211	22.6	20.570	24.6	23.196
18.7	16.196	20.7	18.323	22.7	20.695	24.7	23.335
18.8	16.297	20.8	18.436	22.8	20.820	24.8	23.475
18.9	16.399	20.9	18.549	22.9	20.946	24.9	23.615
19.0	16.501	21.0	18.663	23.0	21.073	25.0	23.756
19.1	16.603	21.1	18.777	23.1	21.201	25.1	23.898
19.2	16.706	21.2	18.892	23.2	21.329	25.2	24.040
19.3	16.810	21.3	19.008	23.3	21.458	25.3	24.184
19.4	16.914	21.4	19.124	23.4	21.587	25.4	24.328
19.5	17.019	21.5	19.241	23.5	21.717	25.5	24.472
19.6	17.124	21.6	19.359	23.6	21.848	25.6	24.618
19.7	17.231	21.7	19.477	23.7	21.980	25.7	24.765
19.8	17.337	21.8	19.596	23.8	22.112	25.8	24.911
19.9	17.444	21.9	19.715	23.9	22.245	25.9	25.059

PRE-LAB QUESTIONS | EXPERIMENT

Gasimetric Analysis
of a Carbonate | 12

1. Explain how a barometer is used to measure the pressure of the atmosphere.

2. Water vapor varies with the temperature of the water. Explain why this is true.

3. Atmospheric scientists are concerned about the increasing levels of $CO_2(g)$ in the environment. Carbon sequestration is the process of capture and long-term storage of atmospheric $CO_2(g)$. One process involves reacting carbon dioxide with an abundantly available metal oxide such as magnesium oxide to form a stable carbonate. Write an equation that represents the sequestration of $CO_2(g)$ by magnesium oxide. (Label the states of matter for all reactants and products.)

 4. Methane is another greenhouse gas. One pound of methane traps 25 times more heat in the atmosphere than a pound of carbon dioxide. Methane is also the main ingredient in natural gas. If methane can be captured from landfills, it can be burned to produce electricity, run vehicles, or heat buildings. Write an equation that represents the combustion of methane. Consider the reactants and products of the reaction, and comment on the benefits/drawbacks of this proposed use of methane.

Name ______________________ Date ______________________

Instructor ______________________

REPORT SHEET | EXPERIMENT

Gasimetric Analysis of a Carbonate

12

Data Table: Gasimetric Analysis

	Trial 1	Trial 2
Mass of sample	________	________
Volume of gas collected	________	________
Δh (in mm)	________	________
Barometric pressure	________	________
Temperature of the room	________	________
Temperature of water	________	________
Vapor pressure of water	________	________
Partial pressure of CO_2	________	________

Show calculation for partial pressure of CO_2 for trial 2. Include proper units and significant figures.

moles of CO_2 collected (experimental) ________ ________

Show calculation for trial 2.

Theoretical # moles of CO_2 produced   _____

Show calculation.

Use calculations to verify that HCl was added in excess.

Percent error   _____

Show work for your calculation. Use trial 2.

CO_2 gas is partially soluble in water. How will this impact the results of your experiment? Be specific.

Design an experiment:

You may work with your partner on this question and no one else.

Scenario: You work for an environmental waste management firm. A large volume of a white powdery solid is found at a waste dump. The label is torn, but you can decipher the last word of the label. It reads "rbonate." How could you verify if this sample is indeed a carbonate?

Next, design an experiment that might allow you to determine the exact identity of the carbonate. Give as much detail as necessary. Use the reverse side of this sheet or attach another sheet if necessary.

Calorimetry: Heat of Fusion and Specific Heat | 13

The *law of conservation of energy* states that energy can neither be created nor destroyed. Energy can, however, be transferred from one part of a system to another. In a closed system, the heat gained by one object must equal the heat lost by another object.

Do not confuse heat with temperature. *Temperature* is a measure of how cold or warm a substance is in relation to another substance. Temperature is measured in degrees. Heat is a form of thermal energy. All forms of energy can be converted to heat. Heat is measured in calories or joules. A **calorie** *is the amount of heat required to raise the temperature of 1.00 g of liquid water by 1.0 degree Celsius*. This is a very small quantity of heat, so heat is often expressed in kilocalories.

We will apply the law of conservation of energy to complete two experiments. The first will measure the amount of heat used to melt ice. The amount of heat required to melt a substance is called the *heat of fusion*. We will melt a quantity of ice by adding it to warm water.

Since the heat gained by a system is equal to the heat lost, we can use this principle to measure the heat required to melt ice. Heat can be observed as a flow of energy. Heat passes from an object at a higher temperature (water) to an object of lower temperature (ice). The two objects in contact will eventually reach the same temperature at equilibrium.

In general, when a substance undergoes a temperature change, the heat required for the change to take place is given by the equation

$$q = m \times C \times \Delta T$$

where q is the change in heat energy, m is the mass of the substance, C is the specific heat of the substance J/g °C, and ΔT is the difference between the initial temperature and the final temperature. By definition, **specific heat** is the amount of heat per unit mass required to raise the temperature of a substance by one degree Celsius. Specific heat is an intensive property that is specific for each substance.

So, what does this mean? Objects with a high specific heat warm up slowly because it requires a good deal of energy to heat these objects. Water has one of the highest known specific heat capacities of any liquid. It takes a longer time period to heat water than most other liquids. Your car uses water to cool the engine. Your engine reaches high temperatures during gas combustion. The metals in the engine heat up quickly, and water is used to cool the engine. You've probably seen a car with steam coming from the hood. This means the engine has overheated. The cooling system circulates water through pipes in the engine. As the water passes through the hot engine, it absorbs heat, cooling the engine. After the water leaves the engine, it passes through a heat exchanger, or radiator, which transfers the heat from the fluid to the air blowing through the exchanger. This is a very simplified discussion of car maintenance. You can learn more at http://auto.howstuffworks.com/cooling-system.htm.

The principle of specific heat is also used to heat things up! You use it every day when cooking. Many metals have low specific heat values. With a low specific heat value, copper heats up quickly. This allows the food in the pan to heat quickly as well when heat is transferred from the pot to the food inside. Cooking pots are often made of aluminum and/or stainless steel, with specific heat values of 0.897 J/g °C and 0.5 J/g °C, respectively. Some pans have a copper bottom since copper has an even lower specific heat value of 0.385 J/g °C. This allows the pan to heat up even more quickly. You want the body of the pan to heat quickly to speed up the cooking process. The handle, on the other hand, should not heat quickly. The handles are often made of plastic. Plastic is not a good conductor and has a high specific heat capacity, allowing you to touch the handle even though the pan is too hot to touch. When discussing specific heat, you may also hear the term *heat capacity*. Heat capacity is the ratio of the amount of energy absorbed to the associated temperature rise.

Table 13.1 lists the specific heat values for some common metals, elements, and household compounds.

As stated earlier, we will use the known specific heat of water to measure Q, the heat required to melt ice. The heat of fusion of water is 80.0 cal/g. This means 80 calories are required to melt one gram of ice to liquid water. The equation is as follows:

$$H_2O_{(s)} \times heat_{fusion} \rightarrow H_2O_{(l)}$$

Water will be cooled by melting ice. The amount of heat lost by the warm water is equal to the amount of heat required to melt the ice. By applying $q = m \times \Delta T$, we can calculate the heat of fusion of water. We may then compare our calculated value to the known or accepted value of 80.0 cal/g.

$$\text{heat lost by water (in calories)} = g \text{ water} \times 1.00 \text{ cal/g °C} \times \Delta T$$

The first part of the experiment involves using an energy equation that relies on knowing the specific heat of the substance you are measuring.

In Part II of the experiment, we will use calorimetry to calculate the specific heat of a metal. A known metal, zinc, will be heated. Upon heating zinc pellets, the hot pellets will be transferred to a cup of water. The amount of heat absorbed by the water can be measured by the temperature difference. The temperature of the water will rise as the temperature of the hot metal pellets decreases. The heat is transferred from the hot metal to the water.

$$-q(\text{metal}) = +q(\text{water})$$

$$\text{mass (metal)} \times C \text{ (metal)} \times \Delta T \text{ of metal} = \text{mass (water)} \times C \text{ (water)} \times \Delta T \text{ of water}$$

All of the mass and temperature values can be measured during the experiment. The specific heat, C, of water is known to be 1.00 cal/g °C. The equation can be rearranged to solve for the specific heat of the zinc metal.

$$C \text{ (metal)} = \frac{\text{mass (water)} \times C \text{ (water)} \times \Delta T \text{ of water}}{\text{mass (metal)} \times \Delta T \text{ of metal}}$$

For our purposes, we will use a Styrofoam cup to construct a calorimeter. A calorimeter is an insulated vessel used to measure heat. By using an insulated container, we will attempt to minimize the exchange of the contents of the calorimeter and its surroundings. In effect, we will minimize the heat lost to the atmosphere by using an insulated container.

We will also assume that no heat is lost to the atmosphere, though we know that a negligible loss occurs. For our calculations, we will assume our calorimeter is efficient, with no heat lost or gained from the system.

In the two experiments, heat is being transferred from the object with the higher temperature to the object at the lower temperature. In Part I, the heat from the warm water will be transferred to the ice, causing the ice to melt. In Part II, the water will gain heat transferred by the zinc pellets. In both cases, heat is transferred.

Table 13.1 Specific Heat Values for Common Substances[1]

Substance	Specific Heat (cal/g °C)
Lead (Pb)	0.038
Tin (Sn)	0.052
Silver (Ag)	0.056
Zinc (Zn)	0.093
Iron (Fe)	0.11
Glass	0.12
Table salt (NaCl)	0.21
Aluminum	0.22
Ethyl alcohol	0.59
Water	1.00

Procedure for Part I: Heat of Fusion of Water

We will assemble a calorimeter made from a Styrofoam cup. Place one cup inside the other for double insulation. Next, place the plastic lid on the top cup. A square of cardboard should be placed on top of the lid. A thermometer will be inserted through the hole in the cardboard and the hole in the lid. This will allow temperature measurements during the experiment. You may insert a digital thermometer probe or use a conventional glass thermometer. Your instructor will give you specific instructions. If you use the glass thermometer, you may also use a rubber stopper to hold the thermometer in place. This setup is shown in Figure 13.1.

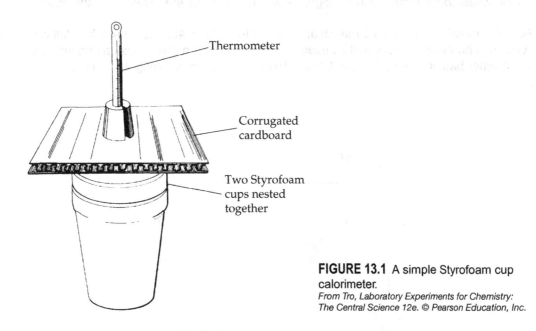

Thermometer

Corrugated cardboard

Two Styrofoam cups nested together

FIGURE 13.1 A simple Styrofoam cup calorimeter.
From Tro, Laboratory Experiments for Chemistry: The Central Science 12e. © Pearson Education, Inc.

Record the mass of the Styrofoam cups. Add 100.0 mL of water to the Styrofoam cup, and reweigh the contents. Record the mass of the cups plus water. Measure the temperature of the water in the calorimeter. Record the value to the tenth place after the decimal.

In a small 50 mL beaker, obtain ice from the ice machine. Be sure the ice is solid. Decant any water that is in the beaker. You should only be adding ice to the calorimeter. No liquid water should be added. You may want to use a paper towel to pat the ice dry before adding it to the calorimeter. Add the ice to the calorimeter and *immediately* replace the lid. Stir the contents of the calorimeter, and watch the temperature drop as the ice melts. When the temperature has stabilized, record the value. (The temperature should be 5 °C or lower.) All the ice should have melted. Now record the mass of the cups and their contents. The increase in mass is due to the ice that has melted.

Calculate the heat of fusion using the data you have collected.

$$\text{heat lost by water} = \text{heat gained by ice}$$

$$\text{heat lost by water} = \text{mass (water)} \times C \text{ (water)} \times \Delta T \text{ of water}$$

$$= \text{heat in calories required to melt ice}$$

Calculate heat of fusion:

$$\text{heat of fusion} = \frac{\text{calories required to melt ice}}{\text{mass of ice}}$$

Calculate the percent error.

$$\text{percent error} = \frac{\text{true value} - \text{experimental value}}{\text{true value}} \times 100$$

Procedure for Part II: Specific Heat of a Metal

Dry the Styrofoam cups from Part I. Record the mass of the cups. Use a graduated cylinder to add 50.0 mL of water to the calorimeter cup. Record the mass of the cups and water. Measure the temperature of the water in the calorimeter.

Obtain a sample of zinc metal pellets. Weigh a 50 mL beaker, and record the mass. Add the metal pellets to the beaker and reweigh. Record the mass of the beaker plus the pellets.

Pour the metal pellets into a large, clean, dry test tube. Fill a 400 mL beaker half-full with water. A Bunsen burner or hotplate will be used to heat the beaker of water. Use a ring stand to set up a hot water bath if using a Bunsen burner. The water should be brought to a boil.

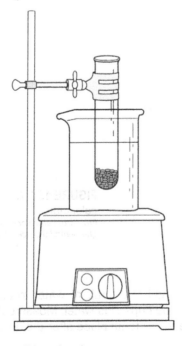

FIGURE 13.2 The metal pellets should be below the surface of the water.
From Timberlake, Laboratory Manual for General, Organic, and Biological Chemistry, 3e. © Pearson Education, Inc.

Add the test tube containing the pellets to the beaker of water. Be careful not to get any water into the test tube. However, be sure to place the metal pellets below the surface of the water as shown in Figure 13.2. It is important that all of the pellets are heated by the water bath. Heat the pellets for a total of 10 minutes.

Next, record the temperature of the boiling water bath used to heat the pellets on the report sheet after the 10-minute heating period. We will assume the pellets reached the temperature of the water bath. Remove the test tube from the water bath. Use a paper towel to wipe the test tube dry. Immediately remove the lid from the calorimeter and add the hot metal pellets. Be careful not to splash water out of the calorimeter. Record the temperature of the calorimeter contents as soon as the lid has been secured in place. Stir the water, and continue to record the temperature of the water every 30 seconds for the next 2 minutes. Record the maximum temperature of the water after the metal pellets have been added.

Calculate the specific heat of the metal using the equation below.

$$C\ (metal) = \frac{mass\ (water) \times C\ (water) \times \Delta T\ of\ water}{mass\ (metal) \times \Delta T\ of\ metal}$$

REFERENCE

1. Bettelheim, F. A., and J. M. Landesberg. 2001. *Laboratory Experiments for General, Organic, and Biochemistry, fourth edition.* Harcourt, Inc.

PRE-LAB QUESTIONS | EXPERIMENT

Calorimetry: Heat of
Fusion and Specific Heat | 13

1. We eat food to gain energy. Some foods provide energy more efficiently than others. The established energy value for three food types is given: carbohydrates, 17 kJ/g; fats, 38 kJ/g; and proteins, 17 kJ/g. Look at the nutrition label on your favorite candy bar. Calculate the energy gained by eating this candy. (Many companies provide nutritional facts online.)

2. If the same amount of heat is added to equal masses of aluminum, lead, zinc, and iron, which metal would experience the greatest increase in temperature? Explain.

3. Styrofoam is a great insulator but a very controversial product. Research some of the environmental and health concerns related to Styrofoam. Cite your sources.

4. Discuss some modern alternatives to Styrofoam. Put the information in your own words, but cite your sources.

Name _____ Date _____

Instructor _____

REPORT SHEET | EXPERIMENT

Calorimetry: Heat of Fusion and Specific Heat | 13

Part I: Heat of Fusion

Mass of empty calorimeter cup   _____ grams

Mass of calorimeter cup + water   _____ grams

Mass of water   _____ grams

Initial water temperature   _____ °C

Final water temperature   _____ °C

ΔT of water   _____ °C

Calories of heat lost by liquid water (neg. value)   _____ calories

Calories of heat required to melt ice (pos. value)   _____ calories

Show calculations.

Mass of calorimeter cup + water + melted ice   _____ grams

Mass of ice that melted   _____ grams

Heat of fusion   _____ calories/g

Show calculations.

Percent error   _____ %

Show calculations.

Part II: Specific Heat

Mass of empty calorimeter cup	_____ grams
Mass of calorimeter cup + water	_____ grams
Mass of water	_____ grams
Mass of 50 mL beaker	_____ grams
Mass of 50 mL beaker + metal pellets	_____ grams
Mass of metal pellets	_____ grams
Initial temperature of water	_____ °C
Temperature of hot metal (boiling water)	_____ °C
ΔT of cold water in calorimeter	_____ °C
Maximum temperature of water after metal is added	_____ °C
Heat gained by water	_____ calories

Show calculations.

Heat lost by metal	_____ calories
Temperature of metal	_____ °C
Specific Heat of metal	_____ cal/g °C

Show calculations.

Percent error   _____ %

Show calculations.

Explain limitations of your calorimeter.

How could you overcome these limitations?

Design an experiment to calculate the energy contained in 5 grams of potato chips.

Chromatography: Isolation and Characterization of Yellow Dye No. 5 | 14

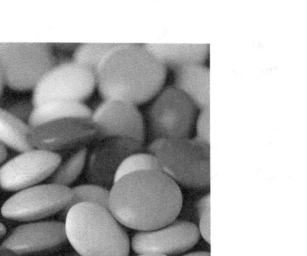

Mark Matysiak/Fotolia

Food and cosmetic products contain a multitude of additives and preservatives. Such substances are used as colorants, emulsifiers, stabilizers, and flavor enhancers. These additives are usually a very small component of foods. Some individuals have adverse reactions to the coloring added to foods. In some cases, the medical community refers to these adverse reactions as allergies. In other cases, they're considered food intolerance.[1]

Color additives are also used in the manufacture of clothing, plastics, paints, and pharmaceuticals. The Food and Drug Administration (FDA) has two major classifications for color additives: (1) colors certified by batch, which are derived primarily from petroleum and coal sources, and (2) colors exempt from batch certification, which are obtained from plant, animal, or mineral sources.[2] Basically, they are divided into natural and synthetic color additives.

Currently, nine color additives are certifiable in the United States. However, only seven are used in food manufacturing. The seven dyes used to color foods are shown in Figure 14.1. Each batch of certified color additive undergoes testing by the manufacturer and the FDA to meet strict specifications for purity. The color additives that are exempt from certification include pigments derived from natural sources. For example, caramel color is produced commercially by heating carbohydrates and sugar under strictly controlled conditions for use in gravies, sauces, soft drinks, and baked goods. The exempt color additives must also meet certain legal standards for stability and purity.

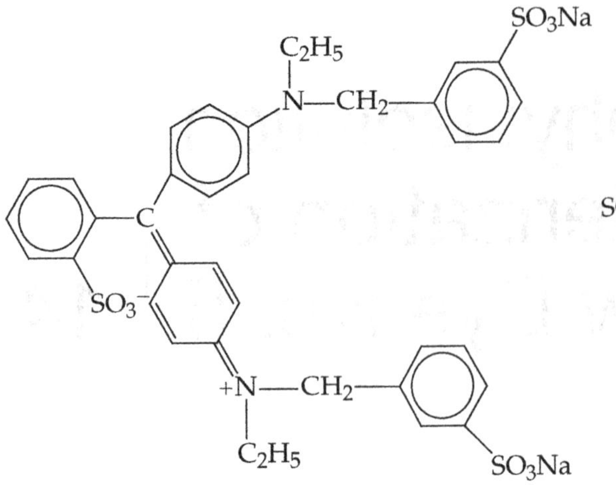

Blue No. 1

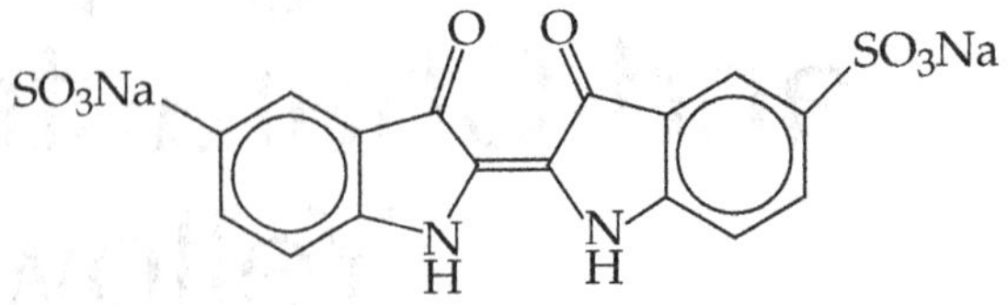

Blue No. 2

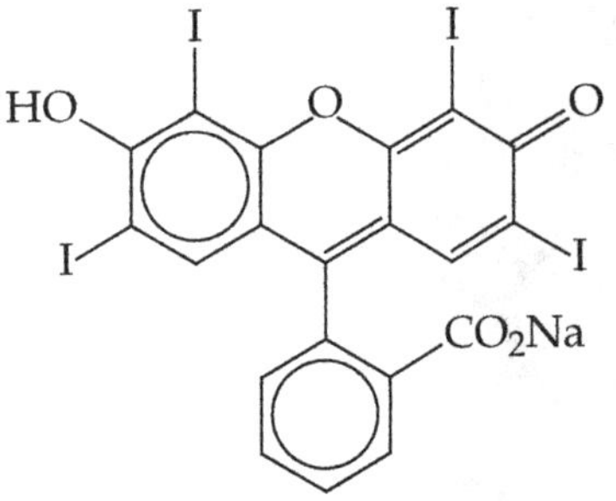

Red No. 3

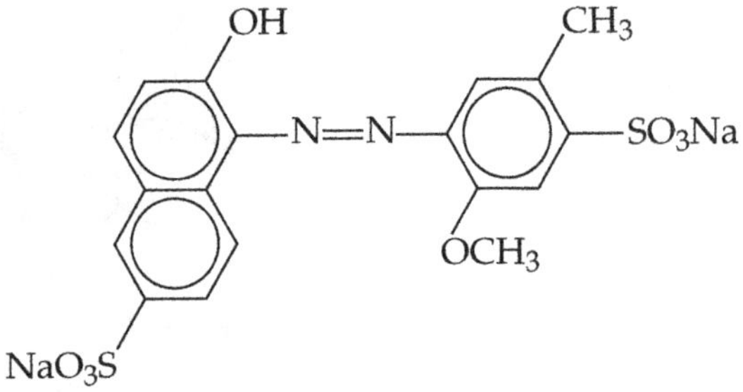

Red No. 40

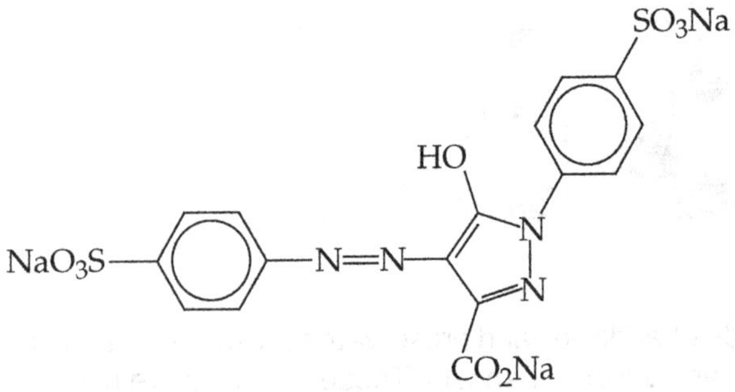

Yellow No. 5

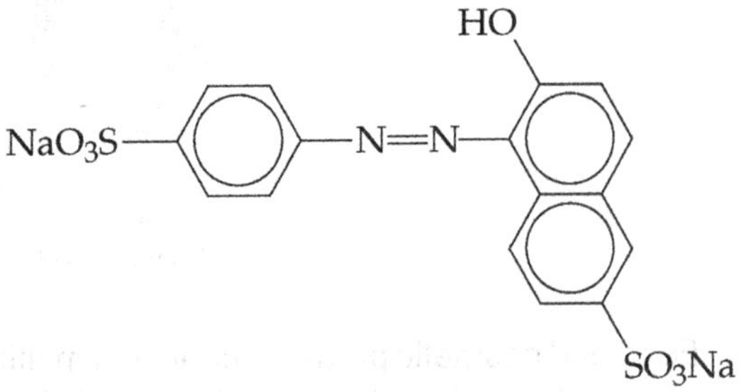

Yellow No. 6

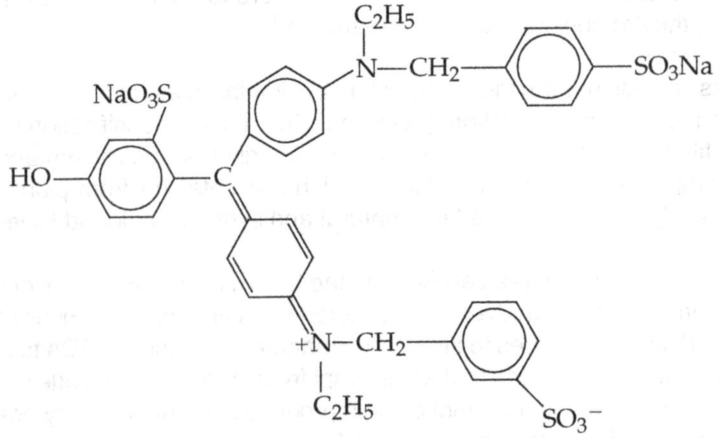

Green No. 3

FIGURE 14.1 Structures for seven dyes found in M&M™ candies.

During this experiment, you will set out to solve a practical problem. Although some additives do not pose any harm or danger to the overall population, individuals may develop an allergic reaction to specific additives. Such is the case with yellow dye No. 5. (Tartrazine is the chemical name for yellow dye No. 5.)

Most food substances contain a mixture of ingredients and color additives. The list of certified color dyes does not include every color found in many candy and food products. Colors such as purple, orange, or pink are produced with a blend of colorants. Often, a mixture of several dyes will be present in one product in order to generate the desired color effect. This lab experiment will focus on the separation of color dye mixtures by paper chromatography.

Chromatography is a technique that involves two phases: a stationary phase and a mobile phase. We will use paper as our stationary phase and a dilute salt solution as our mobile phase. Compounds that prefer the stationary phase move slowly up the paper. Those that prefer the mobile phase move more quickly as the solution moves up the paper. The components of the mixture will separate based on their affinity for the mobile phase. Thus, they will be separated based on solubility in the mobile phase. Once the mixture separates into individual components, the components must be detected. In the case of food dyes, we can detect these components visually by color analysis. Sometimes the analysis is not so straightforward and requires detection by ultraviolet absorption, conductivity, and/or radioactivity measurements.

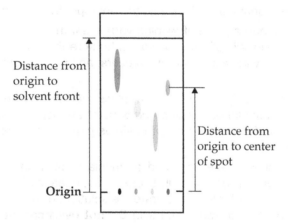

FIGURE 14.2 A pencil is used to mark the origin and the solvent front.

To analyze a mixture with paper chromatography, a small amount of sample is placed on the paper. The site where the sample is introduced is called the origin, as illustrated in Figure 14.2. The samples travel with the mobile phase. The solvent front is the total distance the solvent travels. The retention factor, or R_f value, is the ratio of the distance the substance moved to the distance the solvent moved. An R_f value is calculated for each spot on the chromatogram. It can be stated simply:

$$R_f = \frac{\text{distance traveled by spot}}{\text{distance traveled by solvent}}$$

A Practical Application of Chromatography

Your friend has an allergic reaction to yellow dye No. 5. The M&M package states that the color additive is present in the candy. However, is the dye present in all of the candy coatings, or is it present only in candies of certain colors? Which of the candies, if any, are safe for your friend to consume? You will use paper chromatography to answer this question.

The label on a package of M&M candies lists the following ingredients: milk chocolate (sugar, chocolate, skim milk, cocoa butter, lactose, milk fat, soy lecithin, salt, artificial flavor), sugar, corn-starch, less than 1% of the following: corn syrup, dextrin, coloring (includes blue lake 1, yellow 6, red 40, yellow 5, blue 1, red 40 lake, blue 2 lake, yellow 6 lake, blue 2), and gum acacia.

First, we will extract the dyes from the M&M candy.[3,4] We must be careful to only remove the outer candy coating that covers the chocolate.

PROCEDURE

Extraction of the Dyes

1. First, cut seven 3 cm strips of woolen yarn. Place the yarn in a large test tube containing acetic acid. Heat the test tube in a boiling water bath for 4–5 minutes. (This should remove any fluorescent dyes that could interfere with the separation of the dyes.) Cool to room temperature, and remove the yarn from the test tube. Discard the acetic acid. It may be poured down the sink.

2. Next, place each colored M&M into a separate test tube. Carefully label each tube with a marker. Add approximately 3 mL of acetic acid to each test tube to cover the candies. Heat each test tube in the boiling water bath. Be careful to dissolve only the outside candy coating.

3. Prepare a clean set of test tubes, one for each color candy. Decant each solution, which now contains dyes, sugars, and fillers, into the appropriate test tube.

4. To extract the dyes, add a piece of woolen yarn. Heat this tube in the boiling water bath for 5 minutes. Occasionally stir the solution to be sure the yarn is submerged in the dye solution. Remove the yarn with a pair of tweezers, and rinse it with a small amount of tap water.

5. To release the extracted dye from the yarn, place the "dyed" yarn into a clean test tube. Prepare a separate tube for each color. Add 5 mL of ammonia solution. Mix with a stirring rod. Test the solution for basicity using red litmus paper. If the solution is not basic, add additional ammonia solution.

6. Heat the tubes containing the yarn and ammonia solution in a boiling water bath for 5 minutes. Stir the contents occasionally to ensure the release of the dye. Remove the yarn, and continue to heat in order to concentrate the solution. (If all of the solvent evaporates, add a few drops of distilled water and stir.) **Do not overheat the solution or you can decompose the material.**

Chromatographic Separation

Now that the dyes have been extracted from the candy coating, we must separate the components of each dye. We will use tartrazine as our reference standard for yellow dye No. 5.

1. Cut two 10 cm × 20 cm pieces of chromatography paper. Using a pencil, draw a straight line 1 cm from the long edge of the paper.

2. Next, use a capillary tube to place a small spot of each concentrated dye solution on the pencil line. Evenly space all of the colored spots. Allow the spots to dry, and repeat the markings on each exact spot to increase the color concentration. You may have to spot each color 3–4 times.

3. Next, you must mark a standard for comparison. Weigh approximately 10 mg of tartrazine. Transfer the powdered compound into a test tube. Add 5 mL of distilled water. Stir. Now, spot the standard on your chromatography paper. Follow the same procedure as spotting your unknown dye samples.

4. Prepare a 100 mL solution of 1% m/m NaCl in distilled water. Carefully mix the solution.

5. Transfer 20–30 mL of the solution to a 600 mL beaker. Cover the beaker with aluminum foil or parafilm.

6. Carefully staple the chromatography paper into a cylinder. Do not let the edges of the paper overlap. Place the paper cylinder in the beaker as illustrated in Figure 14.3. Cover with aluminum foil. Do not let the solvent level rise above the line of spots.

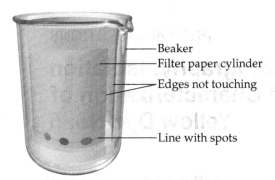

Beaker
Filter paper cylinder
Edges not touching
Line with spots

FIGURE 14.3 Place paper cylinder in the beaker and then cover with aluminum foil or parafilm.
From Tro, Laboratory Manual for Chemistry: A Molecular Approach 3e. © Pearson Education, Inc.

7. Allow the chromatogram to develop until the developing solution reaches 1.5 cm from the top of the paper. Remove the paper, and quickly mark the solvent front with a pencil. Set the paper on a paper towel to dry.

8. Repeat the chromatography separation starting with step (2).

9. Calculate the R_f value for each spot on your chromatogram.

REFERENCES

1. http://www.mayoclinic.com/health/food-allergy/AN01109
2. "FDA/CFSAN Food Color Facts." http://vm.cfsan.fda.gov/list.html.
3. McKone, H. T. 1987. *Laboratory Manual for General, Organic, and Biochemistry.* Ed. William H. Brown and Elizabeth P. Rogers. Moneterey: Brooks/Cole Publishing Company: 159–165.
4. Kandel, M. 1992. *Journal of Chemical Education, 69*: 988.

PRE-LAB QUESTIONS | EXPERIMENT

Chromatography: Isolation and Characterization of Yellow Dye No. 5 | **14**

1. You must prepare 100 mL of a 1% m/m NaCl solution in water. How many grams of NaCl are needed for the solution?

2. Why will you use a pencil to mark the origin of your samples?

3. Use a ruler to calculate the R_f value for A, B, and C.

REPORT SHEET | EXPERIMENT

Chromatography: Isolation and Characterization of Yellow Dye No. 5

14

1. Calculate the R_f value for each spot on your chromatogram. Attach your chromatogram as supporting data.

2. Prepare a table listing each candy and the components separated using chromatography. Include the R_f values.

3. Which M&M candies are safe for your friend to eat? Use your data and calculations to support your answer.

Freezing Point Depression or a Lesson in Making Ice Cream | 15

unpict/Fotolia

"I scream! You scream! We all scream for ice cream!" People rarely get excited about colligative properties, but they should. Colligative properties help us make ice cream! Freezing point depression is a colligative property. A colligative property of a solution depends on the ratio of the number of solute particles to the number of solvent molecules in a solution, and not on the type of chemical species present.

Let's consider how freezing point depression aids in the preparation of ice cream. You can quickly freeze your own ice cream with some salt and ice. (Don't believe me? Follow along at home. The ice cream recipe is listed in the references.)[1] Place the ingredients (milk, sugar, cream) in a quart Ziploc™ bag. Seal the bag. Fill a larger bag (gallon size Ziploc™) with 2 cups of ice and ½ cup of sodium chloride (table salt). Place the sealed bag of ingredients inside the larger bag with ice and salt. Gently rock the bag back and forth. You may want to wear gloves or place a towel around the bag. Continue to rock the bag for 10–15 minutes. Voila! You now have ice cream.

What is happening on the molecular level? Basically, you are freezing the liquid ice cream ingredients. Ice cream freezes at a temperature lower than the freezing point of water. The sugar and fats in the mix interfere with the formation of ice crystals, requiring a lower (colder) temperature to get the ice cream base to harden (freeze). Therefore, you can't use plain ice to chill the ice cream base. The ice will melt before the base gets cold enough. By adding salt to the ice, you lower the freezing point of the ice. The ice/salt mixture reaches a temperature low enough to freeze the ice cream base.

In this experiment, we will determine the molar mass of isopropyl alcohol by investigating the freezing point depression observed in an aqueous solution containing the alcohol. *Colligative properties* depend only on the number, not on the identity, of the solute particles in a solution. We discussed freezing point depression. Other colligative properties include osmotic pressure and boiling point elevation.

Freezing point is the temperature at which the solvent in solution and the pure solid solvent have the same vapor pressure. Basically, the solid and liquid are coexisting. When a solute is added, the vapor pressure of the solvent changes and causes the vapor pressure of the liquid solvent to be lowered. The solute particles are interfering or standing between the solvent particles. This causes intermolecular forces to be weakened. Therefore, lower temperatures are required to make it possible for the solvent particles to reach each other and form the solid.

The freezing point of a solvent is lowered when any solute is dissolved in that solvent. The nature of the solute is not important, only the number of particles of the solute. We can quantify the freezing point depression because it is proportional to the amount of solute present in a given mass of solvent. The molality of the solute is needed to calculate the freezing point depression. *Molality (m)* is defined as the moles of solute per kilogram of solvent. In our case, it will be the moles of isopropyl alcohol divided by the mass of water in kg.

$$m = \text{moles solute/kg solvent}$$

The freezing point depression ΔT_f is a measure of the difference between the freezing point of the solution and the freezing point of the pure solvent. In our case, the pure solvent is water.

$$\Delta T_f = T_{f\ solution} - T_{f\ solvent}$$

The freezing point depression is related to molality and the freezing point depression constant, K_f. The van't Hoff factor, i, indicates how many particles a given solute breaks apart into when dissolved in solution. For nonelectrolytes that dissolve in water, the value is 1. The van't Hoff factor for isopropanol is 1. It is a molecular compound that does not separate into ions. The value of K_f is characteristic for a solvent. The K_f value for water is $-1.86\ °C/m$.

$$\Delta T_f = K_f\, mi$$

You can experimentally determine the molar mass of a solute by measuring the freezing point depression and rearranging the equation.

$$m = \Delta T_f / K_f$$

$$m\ (\text{kg solvent}) = \text{moles solute}$$

$$(\text{mass solute})/(\text{moles solute}) = \text{molar mass of solute}$$

Procedure

Set up a freezing point apparatus as demonstrated by your instructor. The apparatus consists of a large test tube, a two-hole stopper with wire stirrer, and a digital thermometer as seen in Figure 15.1. *Be careful not to push the stopper too far into the test tube.*

Prepare a water/ice/rock salt bath by adding equal amounts of ice and rock salt and a small amount of water in the 1000 mL beaker. Stir the mixture, and check the temperature of the bath. It should be at $-10\ °C$. If the bath is too warm, you may need to decant some water or add more ice/rock salt.

Measure 25.0 mL of distilled water with a graduated cylinder. Transfer the water to the test tube of the freezing point depression apparatus.

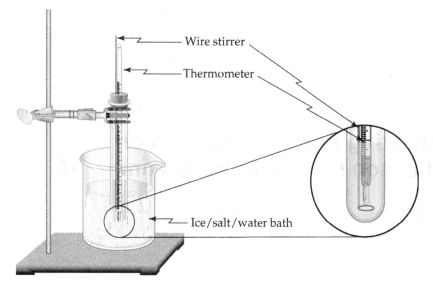

FIGURE 15.1 The wire stirrer will be used to determine if the solution is frozen.
From Tro, Laboratory Experiments for Chemistry: The Central Science 12e. © Pearson Education, Inc.

Use the stirrer to agitate the water in the test tube. Record the time and temperature at 30-second intervals. Continue recording the temperature until the temperature value has been constant for five consecutive readings. (If the water has frozen sufficiently to impair stirring, record the temperature.) Allow the frozen water to thaw, and repeat the procedure a second time.

Allow the frozen water to thaw, and prepare for the next experiment. Add 6–7 mL of isopropyl alcohol into the test tube. Use a buret to dispense an exact volume of alcohol. The volume should be recorded to the hundredth place. Stopper the test tube, and place it in the bath. Again, stir the solution and record the temperature at 30-second intervals. Continue until the readings are the same for five consecutive measurements or the solution has become difficult to stir. Allow the frozen mixture to thaw, and repeat the procedure.

REFERENCE

1. Ice cream recipe: 1/2 cup milk, 1/2 cup whipping cream (heavy cream), 1/4 cup sugar, 1/4 teaspoon vanilla or vanilla flavoring

PRE-LAB QUESTIONS | EXPERIMENT

Freezing Point Depression or a Lesson in Making Ice Cream

15

1. Explain the difference between molarity and molality.

2. Explain why salt is placed on icy sidewalks in the winter.

3. A solution is prepared by mixing 3.62 g of an unknown electrolyte with 285.0 g of chloroform. (Assume the van't Hoff factor is 1.) The freezing point of the solution is −64.2 °C. (The normal freezing point for chloroform is −63.5 °C.) The K_f value for chloroform is −4.68 °C/m. Find the molar mass of the unknown.

Name ______________________________ Date ______________________

Instructor ______________________________

REPORT SHEET | EXPERIMENT

Freezing Point Depression or a Lesson in Making Ice Cream

15

Experimental Data

Density of isopropyl alcohol 0.785 g/mL

	Trial 1	Trial 2
Final buret reading	________	________
Initial buret reading	________	________
Volume of isopropyl alcohol	________	________
Mass of isopropyl alcohol used	________	________
t_f, water	________	________
t_f, mixture	________	________
Δt_f	________	________
molality (experimental)	________	________
moles isopropyl alcohol in sample	________	________
molar mass of isopropyl alcohol	________	________

Show *molality* calculations for one trial. Include proper units and significant figures.

Show the calculations for the number of moles of isopropyl alcohol in the sample.

Percent Error. Calculate the molar mass of isopropyl alcohol, and report the percent error between the experimental value and the literature value.

Alternative Fuel Project | 16

ALTERNATIVE FUEL PROJECT

The United States is the single largest consumer of fossil fuels in the world. According to the U.S. Energy Information Administration, the United States consumed approximately 140 billion gallons of gasoline in 2015, equaling a daily average of 9.16 million barrels. This is a decrease from the 142 billion gallons consumed in 2007. To continue to reduce our consumption, we must seek alternative sources of fuel. To better understand alternative fuel sources, we will investigate current options and proposed solutions. Many college campuses seek to reduce pollution and conserve energy by exploring many transportation alternatives. Some of these options include electric service vehicles, natural gas-powered shuttles, and a flex-car system. To consider alternative fuel sources, we will prepare biodiesel. To better understand biodiesel, we will need to put it in the context of currently existing fuels and other fuel alternatives. Our class will study fuel sources that are currently available and materials that have potential as alternative fuels.

Part I: Instructions

You will work in a group of four students to prepare a response to a series of questions. You may select the members of your group, or your instructor may assign teams. Each group is limited to four participants. The questions you will answer relate to traditional and alternative fuel sources. Your group will become the class experts on this particular topic. It is your goal to convey knowledge to your classmates through a presentation or video.

You are encouraged to be creative in your presentation. You may prepare a video clip, a PowerPoint presentation, a podcast, or other media presentation. Creativity is encouraged, and you are welcome to interview experts in the field. However, the main goal is to educate your classmates about your assigned topic.

A sample Evaluation Form is provided to help you understand the goals of your presentation. This form evaluates your group's performance during the presentation.

At the end of the project, you will complete a group evaluation form describing the quality of work produced by your teammates. (If a group member does not make significant contributions to the group assignment, the instructor may choose to adjust individual scores.) Overall, students work well in groups and learn from each other. The evaluation form is a safeguard to prevent an unbalanced workload.

Selecting Topics

A series of seven questions has been developed for the course project. Groups will sign up for the topics on a first-come, first-served basis.

Questions

1. How does a traditional internal combustion engine work? How does a diesel engine work? What is a spark plug?

2. How is gasoline prepared? How have industrial procedures changed over time, and what additives have been used and removed?

3. How is diesel fuel prepared? Petroleum diesel? Synthetic diesel? Biodiesel? New routes to diesel?

4. Why is ethanol being considered as an alternative fuel? What are the benefits and drawbacks to using ethanol? What about methanol?

5. Why is hydrogen being considered as an alternative fuel? What are the benefits and drawbacks to using hydrogen? How does a fuel cell work?

6. Many vehicles run on natural gas. What are the advantages and disadvantages of this fuel source? What is the difference between compressed natural gas (CNG) and liquefied natural gas (LNG)?

7. Electric vehicles have entered the mainstream market. They typically require charging stations. How does an electric vehicle work? What are the advantages and disadvantages of an electric vehicle?

ONLINE RESOURCES FOR THE ALTERNATIVE FUEL PROJECT

This is not an exhaustive list of resources. This is simply a short list of sources to help you get started with your research. You may also find reference books in the library.

http://energy.gov/eere/office-energy-efficiency-renewable-energy

Glossary of terms commonly used in energy literature. It contains a huge list, and the terms are all very well defined. You should explain the terms you use during your presentation if your audience might not know them.

http://www.energyquest.ca.gov/transportation/

Website for students created by California's Energy Commission that has a short description of all the alternative fuels.

http://www.eere.energy.gov/afdc/fuels/biodiesel.html

The Department of Energy website that has more in-depth information on all of the alternative fuels plus information on their physical properties, cost, and U.S. distribution.

http://auto.howstuffworks.com/

A website devoted to how automobile engines, including electric cars, work.

http://www.irvingoil.com/company/refinery.asp

An oil refinery's website that has a really nice interactive tutorial on the refining process.

http://www.hydrogen.energy.gov/

A site that gives an overview of hydrogen with each step of the process (production, delivery, storage, use, etc.).

http://www.eia.doe.gov/basics/naturalgas_basics.html

A site that serves as a good introduction to natural gas.

http://www.naturalgas.org/

A website that has some good information on natural gas but is very one-sided because it is run by natural gas companies.

Alternative Fuel Presentation

Group: _____

Use the following categories to assign numerical scores from 0 to 10. Be fair and consistent. Tally the scores.

	Score		Score		Score		Score
Outstanding	10	Above Average	8	Below Average	5 to 6	Unsatisfactory	1 to 2
Excellent	9	Average	7	Poor	3 to 4	Did not participate	0

Evaluation Criteria	Score
1. Oral presentation was clear/good interactions with audience.	
2. Presentation answered assigned questions completely.	
3. Visual aids were appropriate and effective.	
4. Ability to keep audience's attention.	
5. Creativity of presentation.	
6. Quality of references/resources.	
7. Thoroughness of research and supporting evidence.	
8. Arguments were logical and well presented.	
9. Ability to inform the audience. Did group *know* the material?	
10. Ability to respond to questions.	
TOTAL -	

Constructive Comments:

PART II: PREPARATION OF BIODIESEL

$$\underset{\text{Triglyceride}}{\begin{array}{c} CH_2{-}O{-}C\overset{O}{\diagup}R \\ | \\ CH{-}O{-}C\overset{O}{\diagup}R \\ | \\ CH_2{-}O{-}C\overset{O}{\diagup}R \end{array}} + \underset{\text{Methanol}}{CH_3OH} \; \underset{}{\overset{NaOH}{\rightleftharpoons}} \; \underset{\text{Biodiesel}}{3\,CH_3{-}O{-}C\overset{O}{\diagup}R} + \underset{\text{Glycerol}}{\begin{array}{c} OH \\ | \\ {-}OH \\ | \\ OH \end{array}}$$

You have studied a series of alternative energy sources. One of these fuels, biodiesel, can be easily prepared in the laboratory. We will use household vegetable oils as our source of triglycerides. Place 100 mL of vegetable oil in a 250 mL beaker. Place the beaker on a hotplate, and heat the oil to 60 °C to speed up the reaction. *However, be careful to not let the temperature exceed 70 °C.*

Add 0.35 g of anhydrous NaOH into a 250 mL Erlenmeyer flask. Add ~20 mL of methanol to the same flask, and place the flask on a stirring plate with a magnetic stir bar. This will dissolve the solid NaOH and create sodium methoxide ($NaOCH_3$), which serves as our catalyst. Once the NaOH has thoroughly dissolved, pour the heated vegetable oil into the flask. Maintain the reaction at 60 °C, and continue stirring for 30 minutes. *Caution: NaOH is highly caustic. If it comes in contact with your skin, flush with water and contact your instructor immediately.*

Transfer the reaction mixture to a separatory funnel. The two products will separate based on density. The denser component, glycerol, will settle to the bottom of the funnel and the biodiesel will float in the top layer. The physical separation is a slow process that will require a minimum of one hour. You may work on your proposal while the mixture separates. See Part III.

Once the mixture has completely separated, drain the bottom layer (glycerol) into a clean beaker and place in the collection bottle labeled glycerol. Use another clean beaker to collect the biodiesel and place in the appropriate bottle.

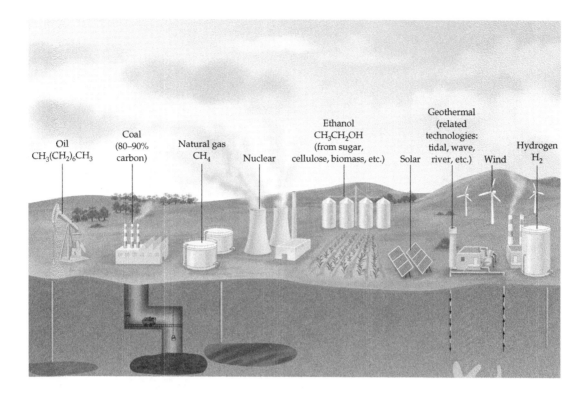

ALTERNATIVE FUEL PROJECT PART III

Many cities in the United States face serious problems with traffic congestion and do not have the infrastructure in place to support a metrowide public transit system. Many nonprofit and government agencies try to minimize these problems by providing incentives to use public transportation in addition to coordinating car pools, shuttles, and flex-car systems.

Last week, you had the opportunity to learn about a range of fuel alternatives. Using that knowledge, your group should respond to the following case study. One typewritten response should be submitted for your group. Your writing should be concise, yet thoroughly address your points. Use references to strengthen your argument. Your proposal will be due at the end of the period. Proof your proposal carefully to generate a professional piece of work free of errors.

Case Scenario:

The Campus Office of Sustainability has received a $2.3 million grant from a private foundation. The funds are restricted for development of an alternative transportation system for the surrounding areas. The university has called for proposals that address energy issues by devising a plan of action.

Develop a five-year action plan for implementing a new transportation system. Be specific as to how the plan would be placed in action. Discuss how the new plan will build on or transform existing plans. Indicate how you will overcome obstacles associated with your plan.

The Green Fades Away | 17

The Rate of Decolorization of a Dye

This experiment will study the kinetics of a reaction involving a green dye. Upon reacting with a Lewis base, the dye undergoes structural changes that cause the molecule in solution to convert from a green color to colorless.

Before measuring how quickly the color disappears, let's consider why the molecule has color. Dyes are aromatic organic compounds. An *aromatic compound* contains planar ring systems with alternating double bonds. The conjugated compound gains chemical stability from the delocalization of electrons in the ring system, as seen in benzene. The electrons are not associated with a single atom or one covalent bond. Instead, delocalized pi electrons form an orbital cloud that extends over several adjacent atoms. Specifically, a *chromophore* is the part of a molecule responsible for its color. The color arises when a molecule absorbs certain wavelengths of visible light and transmits or reflects others. The chromophore is a region in the molecule where the energy difference between two different molecular orbitals falls within the range of the visible spectrum. Visible light that hits the chromophore can thus be absorbed by exciting an electron from its ground state into an excited state. In the conjugated chromophores, the electrons jump between energy levels that are extended pi orbitals, created by a series of alternating single and double bonds.[1]

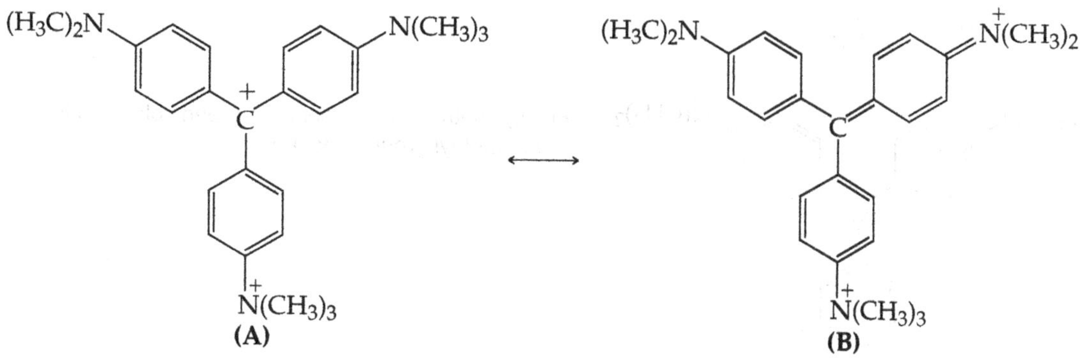

FIGURE 17.1 Two resonance structures of methyl green.

The parent dye does not contain any methyl groups. This structure is called *pararosanline* and is a red color.

When four methyl groups are added, the dye appears a reddish purple, as seen in *methyl violet*.

As more methyl groups are added, the color appears blue, as is the case in crystal violet with six methyl groups.

Finally, methyl green displays a green color with its seven methyl groups attached.

The structure of methyl green can be illustrated with resonance structures. As seen in Figure 17.1 (A), the central carbon atom can bear a positive charge. It can act as a Lewis acid. This means it will act as an electron pair acceptor and react with a Lewis base to form an adduct. The carbon atom will share the electron pair furnished by the Lewis base.

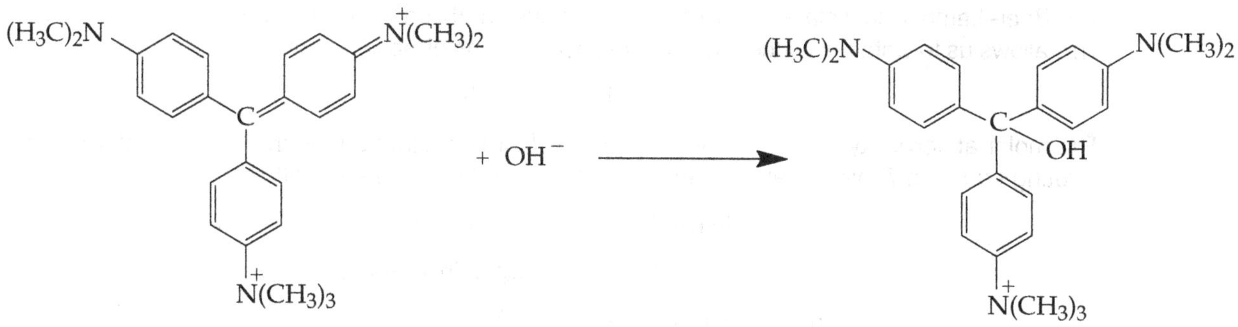

FIGURE 17.2 The reaction of methyl green with hydroxide ion.

When a Lewis base such as a hydroxide ion adds to the central carbon in methyl green, the central atom will now have four single bonds, as shown in Figure 17.2. The geometry around the central atom converts from trigonal planar to tetrahedral. The tetrahedral geometry does not allow pi bonding. The color of the dye will disappear because electron delocalization no longer exists over the entire molecule.

The color of the methyl green solution fades under basic conditions. The rate of the reaction is faster as the conditions become more basic. In the experiment, we will mix solutions of 2.0×10^{-4} M methyl green with basic buffer solutions. We will use 0.040 M aqueous solutions of a Na_2HPO_4 and Na_3PO_4 buffer system. The buffer generates an OH^- ion that reacts with methyl green.

$$PO_4{}^{3-} + H_2O \rightleftharpoons HPO_4{}^{2-} + OH^-$$

The equilibrium expression can be written:

$$K_b = \frac{[HPO_4{}^{2-}][OH^-]}{[PO_4{}^{3-}]} = 4.5 \times 10^{-2}$$

By rearranging the equation, you can determine the concentration of OH^-. It is important to note that the actual concentration of the buffer ions is not important. It is their relative concentrations that matter.

$$[OH^-] = 4.5 \times 10^{-2} \times \frac{[PO_4{}^{3-}]}{[HPO_4{}^{2-}]}$$

The reaction is first order with respect to the dye. When the concentration of the dye doubles, the rate also doubles. The observed rate law is:

$$\text{rate} = k_1[\text{dye}][H_2O] + k_2[\text{dye}][OH^-]$$

The reaction acts as if it were first order when the concentrations of both H_2O and OH^- are much greater than the concentration of the dye. For our reaction, the concentration of methyl green dye will be 1.0×10^{-4} M. Thus, $[H_2O] \gg [\text{dye}]$ and $[OH^-] \gg [\text{dye}]$. Under these conditions, the concentration of H_2O and OH^- essentially remains constant and the reaction is called pseudo-first order. We can write an observed rate constant that includes the quantities of $[H_2O]$ and $[OH^-]$.

$$k_{obs} = k_1[H_2O] + k_2[OH^-]$$

By substituting the observed rate constant into the original rate equation, we obtain this equation:

$$\text{rate} = k_{obs}[\text{dye}]$$

Integrating the new equation yields:

$$\ln[\text{dye}] = -k_{obs}\,t + \ln[\text{dye}]_o \quad (6)$$

The Beer–Lambert law states the relationships between absorbance and concentration: $A = \varepsilon l\,c$. This allows us to write the concentration of the dye in terms of absorbance:

$$[\text{dye}] = A / \varepsilon l \quad (7)$$

The molar absorptivity, ε, and the cell pathlength, l, are constant during the experiment. By substituting equation 7 into equation 6, the integrated rate law can be rewritten:

$$\ln (A/\varepsilon l\,) = -k_{obs}t + \ln(A/\varepsilon l)$$

or $$\ln A - \ln(\varepsilon l) = -k_{obs}t + \ln A_o - \ln(\varepsilon l)$$

or $$-\ln A = -k_{obs}t - \ln A_o$$

Therefore, we can determine the rate constant of the reaction by plotting the negative value of the natural logarithm of the absorbance, $-\ln A$, versus time. The slope of the graph is the rate constant, k_{obs}.

Procedure

Set the dial on the spectrophotometer to a wavelength of 590 nm.

Table 17.1

Trial	0.040 M Na$_3$PO$_4$ (mL)	0.040 M Na$_2$HPO$_4$ (mL)
1	5.0	10.0
2	7.0	8.0
3	8.0	7.0
4	10.0	5.0

Use a pipet or dispensette pump to add the exact volumes of the 0.040 M Na$_3$PO$_4$ and 0.040 M Na$_2$HPO$_4$ solutions for the first trial listed in Table 17.1. Pipet carefully to add exact amounts. Swirl the contents to mix the solutions. Now, prepare to act quickly because the reaction occurs rapidly. Add 5.0 mL of the methyl green solution. Swirl to mix. Rinse the cuvette by filling it with solution and pouring the contents back into the beaker. Now, fill the cuvette and place it in the spectrophotometer. Be careful to properly align the cuvette in the spectrophotometer.

Immediately start recording time and read the absorbance of the solution. Continue to take an absorbance reading every 15 seconds until the absorbance value drops below 0.20. You should have a minimum of five data points for time versus temperature. Repeat for each trial.

REFERENCE

1. http://stainsfile.info/StainsFile/dyes/dyecolor.htm

PRE-LAB QUESTIONS | EXPERIMENT

The Green Fades Away | 17

1. Anthocyanins are a class of compounds found in many berries and other fruits. Draw the general structure and indicate the structural component responsible for creating color.

2. Give a biological application of methyl green.

3. Define Lewis acid and Lewis base.

4. What does it mean for a reaction to be pseudo-first order? How is it different from a first-order reaction?

REPORT SHEET | EXPERIMENT

The Green Fades Away | 17

Data Trial 1

(5.0 mL 0.040 M Na_3PO_4 + 10.0 mL 0.040 M Na_2HPO_4 + 0.0 mL H_2O)

Time	Absorbance (A)	Seconds from Start	$-\ln A$

Data Trial 2

(7.0 mL 0.040 M Na_3PO_4 + 8.0 mL 0.040 M Na_2HPO_4 + 0.0 mL H_2O)

Time	Absorbance (A)	Seconds from Start	$-\ln A$

Data Trial 3

(8.0 mL 0.040 M Na_3PO_4 + 7.0 mL 0.040 M Na_2HPO_4 + 0.0 mL H_2O)

Time	Absorbance (A)	Seconds from Start	−lnA

Data Trial 4

(10.0 mL 0.040 M Na_3PO_4 + 5.0 mL 0.040 M Na_2HPO_4 + 0.0 mL H_2O)

Time	Absorbance (A)	Seconds from Start	−lnA

DATA ANALYSIS

Plot $-\ln A$ versus time for each trial. Determine the slope of all five data sets. The slopes represent the observed rate constants. Record the data in the table below. Submit all of your $\ln A$ vs. time plots.

Complete the following table.

Trial	$[OH^-]$ (M)	k_{obs} (s^{-1})
1		
2		
3		
4		

Calculate the $[OH^-]$ concentration for each trial and record in the table. Show a sample calculation for $[OH^-]$.

Plot the observed rate constants versus the hydroxide concentrations. The slope of this plot is the second-order rate constant for the reaction with hydroxide ion.

1. Why is the slope of k_{obs} versus $[OH^-]$ equal to k_2?

2. Describe an experiment that would examine the effect of the observed rate constant on an increase in ionic strength.

Chemical Kinetics | 18

Chemical kinetics is an area of chemistry that focuses on reaction rates. The reaction rate of a chemical reaction is defined as the change in concentration of a reactant or product per unit time. The molar concentration of reactant will decrease over time, and the molar concentration of product will increase.

There are many practical applications of kinetics. An important branch of science is pharmacokinetics. This field studies the process by which a drug is absorbed, distributed, metabolized, and eliminated from the body. For example, the mechanisms for the sustained release of drugs are based on the half-life of the substances used and sometimes the pH of the body as well. Practically, this affects the way in which dosages are determined and prescribed.

Chemical kinetics is also used in environmental studies to model the rate of ozone depletion in the atmosphere and to study migration of pesticides in soil systems. It has many industrial applications as well. Engineers study the rate of corrosion on metal and the mechanical behavior of rubber. Many chemical companies have entire teams of scientists devoted to catalysis. **Catalysis** is the increase in rate of a chemical reaction due to the participation of a substance called a **catalyst**. A catalyst is a substance that enables a chemical reaction to proceed at a faster rate or under different conditions (such as at a lower temperature) than otherwise possible. Unlike other reagents in the chemical reaction, a catalyst is not consumed. It is estimated that 90% of all commercially produced chemical products involve a catalyst at some stage in the process of their manufacture.[1]

The rate of a reaction is dependent on the concentration of reactants. (If a catalyst is involved, it is also dependent on the catalyst.) Consider the example shown below.

$$2\ NO(g) + 2\ H_2(g) \rightarrow N_2(g) + 2\ H_2O(g)$$

The reaction rate was experimentally determined to be Rate = k $[NO]^2[H_2]$. The superscripts in this equation indicate the order of reaction for each reactant. The reaction order indicates what specific effect a change in concentration of that reactant will have on the rate. The reaction is second order with respect to NO and first order with respect to H_2. If the concentration of hydrogen is doubled, the overall reaction rate will double. If the concentration of NO is doubled, the overall reaction rate will be quadrupled. The overall rate law is the sum of all exponents. In this case, it is third order.

It is very important to realize that the exponent representing the order of each reactant cannot be determined by the written equation. It must be determined experimentally.

Temperature Dependence of a Reaction Rate: Arrhenius Equation[2]

The temperature dependence of reaction rate is expressed through the rate constant. For most reactions, the rate constant varies with temperature according to the Arrhenius equation:

$$k = Ae^{-Ea/RT}$$

In this equation, e is the base of natural logarithms, R is the gas constant 8.314 J/(K mol), and T is absolute temperature. An increase in temperature will lead to a larger k value and consequently a higher reaction rate. A is the frequency factor, a constant that reflects the frequency of collisions with proper orientation. Finally, Ea is the activation energy for the reaction. The activation energy is the difference in energy between the reactants and the transition state, as

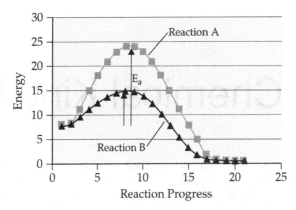

FIGURE 18.1 Reaction A will have a slower rate than Reaction B because it has a greater activation energy.

shown in Figure 18.1. The activation energy is the energy barrier that must be overcome for the reaction to proceed.

Determining a Rate Law

In this experiment, we will determine the rate law and activation energy for the reaction of iron (III) ion with iodide ion to form iron (II) ion and triiodide ion.

$$2 \, Fe^{3+}(aq) + 3 \, I^-(aq) \rightarrow 2 \, Fe^{2+}(aq) + I_3^-(aq)$$

The rate law can be written as Reaction Rate = $k[Fe^{3+}]^m[I^-]^n$.

Our task is to determine the value for the exponents *m* and *n*. This can only be done experimentally. The reaction rate can be written in terms of triiodide ion formation as follows.

$$Rate = \Delta[I_3^-]/\Delta t$$

Unfortunately, there is no practical way to measure the reaction rate directly. However, we can couple the iron (III)–iodide reaction with another reaction and monitor those results.

$$2 \, Fe^{3+}(aq) + 3 \, I^-(aq) \rightarrow 2 \, Fe^{2+}(aq) + I_3^-(aq)$$
$$I_3^-(aq) + 2 \, S_2O_3^{2-}(aq) \rightarrow 3 \, I^-(aq) + S_4O_6^{2-}(aq)$$

During the experiment, $Fe^{3+}(aq)$ is combined with $I^-(aq)$ and $S_2O_3^{2-}(aq)$. The triiodide ion that is formed from the first reaction will react further with the thiosulfate ion present in the reaction mixture. As a result of this coupling, the rate of the iron (III)–iodide reaction can be written in terms of thiosulfate ion as follows.

$$Rate = \Delta[I_3^-]/\Delta t = -1/2\{\Delta[S_2O_3^{2-}]/\Delta t\}$$

By following the change in concentration of thiosulfate ion, the rate of the iron (III)–iodide reaction may be determined.

We can use this coupling reaction because we have a way of monitoring the concentration of thiosulfate ion. Starch is also added to the reaction mixture. Any triiodide ion that does not react with thiosulfate ion will form a dark blue complex with starch.

$$I_3^-(aq) + starch(aq) \rightarrow starch\text{-}I_3^- \text{ complex (blue)}$$

Triiodide ion will react with $S_2O_3^{2-}$ (instead of starch) as long as $S_2O_3^{2-}$ is present. As long as thiosulfate ion remains, the solution will be colorless. When all of the thiosulfate ion has reacted completely, the remaining triiodide ion will be free to form the complex with starch. The solution will turn blue.

In each trial that is carried out during the experiment, the concentration of $S_2O_3^{2-}$ will be 0.00040 M. The time required for complete disappearance of $S_2O_3^{2-}$ is equal to the time required for the solution to turn blue. Consequently, the rate of the iron (III)–iodide reaction is given as

$$\text{Rate} = -1/2\, \{\Delta[S_2O_3^{2-}]/\Delta t\} = -1/2\, \{(0.00040\ \text{M})/\Delta t\}$$

where Δt is the time required for the reaction to turn blue. This is the time interval from mixing reactants to formation of the blue complex.

Determination of the Reaction Order

To determine the exponents in the rate law (i.e., the reaction order), it is necessary to establish how the rate is related to the concentration of each reactant. This requires holding the concentration of one reactant constant while varying the concentration of the other reactant. Any observable change in reaction rate can then be attributed to one particular reactant.

Recall the form of the rate law for the iron (III)–iodide reaction.

$$\text{Rate} = k[Fe^{3+}]^m[I^-]^n$$

Taking the natural logarithm of both sides, the equation becomes

$$\ln(\text{Rate}) = \ln(k) + \ln([Fe^{3+}]^m) + \ln([I^-]^n)$$

or

$$\ln(\text{Rate}) = \ln(k) + m \cdot \ln([Fe^{3+}]) + n \cdot \ln([I^-])$$

If $[I^-]$ is held constant while $[Fe^{3+}]$ is changed, then $n \cdot \ln([I^-])$ is a constant. Since k is also a constant, the two factors can be combined, and the equation becomes

$$\ln(\text{Rate}) = m \cdot \ln([Fe^{3+}]) + \ln(\text{constant})$$

This equation is written in the form of a straight line where $y = \ln(\text{Rate})$ and $x = \ln([Fe^{3+}])$. A plot of $\ln(\text{Rate})$ versus $\ln([Fe^{3+}])$ will give a straight line with slope equal to m, the order of the reaction with respect to Fe^{3+}.

If $[Fe^{3+}]$ is held constant while $[I^-]$ is changed, then $m \cdot \ln([Fe^{3+}])$ is a constant, and the equation becomes

$$\ln(\text{Rate}) = n \cdot \ln([I^-]) + \ln(\text{constant})$$

This equation is written in the form of a straight line where $y = \ln(\text{Rate})$ and $x = \ln([I^-])$. A plot of $\ln(\text{Rate})$ versus $\ln([I^-])$ will give a straight line with slope equal to n, the order of the reaction with respect to I^-.

Determination of the Activation Energy

The most common method for determining Ea for a reaction involves measuring the rate constant, k, at several temperatures. We can calculate the rate constant, k, because the exponents in the rate law have been determined. As the reaction temperature is varied, the value of k will change. The k versus temperature data can be used to determine activation energy using the Arrhenius equation, $k = Ae^{-Ea/RT}$. A linear form of the equation can be derived by taking the natural logarithm as seen below. A plot of $\ln(k)$ versus 1/T will result in a straight line with slope equal to $-Ea/R$.

$$\ln(k) = -\frac{Ea}{R} \cdot \frac{1}{T} + \ln(A)$$

Procedure

Beaker A

Trial#	0.04 M Fe^{3+} (mL)	0.15 M HNO_3 (mL)	H_2O (mL)
1	5.00	10.00	10.00
2	7.50	7.50	10.00
3	10.00	5.00	10.00
4	12.50	2.50	10.00
5	5.00	10.00	10.00
6	5.00	10.00	10.00
7	5.00	10.00	10.00

Beaker B

Trial#	0.04 M KI (mL)	0.004 M $S_2O_3^{2-}$ (mL)	Starch (mL)	H_2O (mL)
1	5.00	5.00	2.50	12.50
2	5.00	5.00	2.50	12.50
3	5.00	5.00	2.50	12.50
4	5.00	5.00	2.50	12.50
5	2.50	5.00	2.50	15.00
6	7.50	5.00	2.50	10.00
7	10.00	5.00	2.50	7.50

Prepare the two solutions for trial 1 shown in the table above. Use pipets to add the appropriate volume of reagents to beaker A and beaker B. The volumes must be exact. Record the temperature of the solutions to 0.1 °C. Add the contents of beaker A to beaker B *and* start a stopwatch timer simultaneously. Swirl the solution to make certain the contents are well mixed. Stop the timer when the reaction mixture reaches a blue color. (The actual shade of blue when the reaction is stopped is subjective. Make certain that whatever the shade of blue used, each trial is taken to that same exact shade of blue. Be consistent.) Record the time. Dispose of the solution by diluting the mixture with water and pouring it down the sink. Clean and dry the beakers. Repeat this process for trials 2 through 7.

Beaker A

Trial#	Temp	0.04 M Fe^{3+} (mL)	0.15 M HNO_3 (mL)	H_2O (mL)
8	RT	5.00	10.00	10.00
9	high	5.00	10.00	10.00
10	low	5.00	10.00	10.00

Beaker B

Trial#	Temp	0.04 M KI (mL)	0.004 M $S_2O_3^{2-}$ (mL)	Starch (mL)	H_2O (mL)
8	RT	5.00	5.00	2.50	12.50
9	high	5.00	5.00	2.50	12.50
10	low	5.00	5.00	2.50	12.50

Next, you will study the reaction at varying temperatures. You have already collected data at room temperature. (This was done in trial 1. The data from trial 1 can be used for trial 8.) The reaction will also be carried out at a temperature approximately 5 °C higher than room temperature. A water bath will be used to maintain the elevated temperature. Your instructor will provide instructions for using the water bath. Add reagents to beaker A and beaker B as indicated in the table. Place the two beakers into the water bath, and allow them to sit for a few minutes to reach temperature equilibrium. Be careful not to allow any mixing of the reaction solution with the water in the bath. The beakers must be monitored at all times. Once the solutions have reached the appropriate temperature, add the contents of beaker A to beaker B and start the timer. Record the temperature of the solution to 0.1 °C. Swirl to mix the contents, and place the beaker back into the water bath. Stop the timer when the reaction mixture reaches a blue color. Record the time, and dispose of the solution.

The reaction must also be carried out at a temperature approximately 5 °C lower than room temperature. Proceed as in trial 9 using a cold water bath.

Waste Disposal

Acidic solutions (0.15 M HNO_3 and 0.04 M $Fe(NO_3)_3$ in 0.15 M HNO_3) should be neutralized, then diluted and poured down the sink. All other solutions should be diluted and poured down the sink.

REFERENCES

1. "Recognizing the Best in Innovation: Breakthrough Catalyst." *R&D Magazine*, September 2005, p. 20.
2. Burkart, Maureen. This experiment was developed in collaboration with faculty from Georgia Perimeter College, Dunwoody Campus.

PRE-LAB QUESTIONS | EXPERIMENT
Chemical Kinetics | 18

1. Consider the following reaction between acetone and bromine.

$$CH_3COCH_3 + Br_2 \xrightarrow{H^+} CH_3COCH_2Br + HBr$$

The following initial rate data were obtained for the reaction at 28.3 °C.

Trial#	[CH$_3$COCH$_3$]	[Br$_2$]	[H$^+$]	Rate (M/s)
1	0.30	0.050	0.050	5.7E-05
2	0.30	0.10	0.050	5.7E-05
3	0.30	0.050	0.10	1.2E-04
4	0.40	0.050	0.20	3.1E-04
5	0.40	0.050	0.050	7.6E-05
*All concentrations are expressed in molarity.				

Determine the rate law for the reaction and the value of the rate constant at 28.3 °C.

Part I: Reaction Rate

Data Table:

Concentration of $S_2O_3^{2-}$ _____

Trial	Initial Temperature °C	Reaction Time (Δt, sec)
1	_____	_____
2	_____	_____
3	_____	_____
4	_____	_____
5	_____	_____
6	_____	_____
7	_____	_____
8	_____	_____
9	_____	_____
10	_____	_____

Trial	Initial Rate	In Rate
1	_____	_____
2	_____	_____
3	_____	_____
4	_____	_____
5	_____	_____
6	_____	_____
7	_____	_____
8	_____	_____
9	_____	_____
10	_____	_____

Part II: Reaction Rate with Respect to $[I^-]$ and $[Fe^{3+}]$

Trial	$[I^-]$	ln $[I^-]$	$[Fe^{3+}]$	ln $[Fe^{3+}]$
1				
2				
3				
4				
5				
6				
7				
8				
9				
10				

Prepare two graphs as per the instructions in the lab handout. The first graph will plot ln(Rate) versus ln of $[Fe^{3+}]$. The second graph will plot ln(Rate) versus ln of $[I^-]$. Attach the graphs to your report sheet.

Using the data from the graphs, write the rate law for the reaction:

Part III: Activation Energy

Use the rate law equation to calculate the rate constant, k, at each of the three temperatures in trials 8, 9, and 10. Use the following equation to prepare a graph of $\ln(k)$ versus $1/T$. Attach the graph, and calculate the activation energy. The activation energy can be determined from the slope of this line, where the slope $= -Ea/R$.

$$\ln(k) = -\frac{Ea}{R} \cdot \frac{1}{T} + \ln(A)$$

Trial	k	ln(k)	1/T
8			
9			
10			

Activation Energy _____

Discuss sources of error for the kinetics experiment. What effect should these sources of error have on the experimental results?

Analysis of Phosphoric Acid in Coca-Cola Classic: Spectrophotometric Analysis | 19

kazoka303030/Fotolia

The Coca-Cola trademark is one of the mostly widely recognized trademarks in the entire world. The success stems from a unique product with highly guarded secret ingredients. Over a century has passed and no one has determined the exact secret formula. The actual recipe is stored in a vault in the World of Coca-Cola Museum in downtown Atlanta, Georgia.[1] Although numerous outlets around the world have a franchise to bottle and distribute Coke, none knows the precise ingredients. They are simply supplied with syrups and other ingredients from The Coca-Cola Company, and they mix them together with carbonated water.

The label on a bottle of Coke states the main ingredients. They include carbonated water, high-fructose corn syrup and/or sucrose, caramel color, phosphoric acid, natural flavors, and caffeine. Of course, this information is vague and does not state the procedure for preparing the beverage. The ratio of reagents and the order in which the materials are mixed control the outcome of the taste.

While some individuals are desperately seeking to crack the secret mix, others are concerned about the beverage contents. The Coca-Cola Company website even responds to the frequently asked question: "Does the acidity in Coke damage teeth and bones?"

Phosphoric acid is deliberately added to soft drinks to give them a sharper flavor. It also slows the growth of molds and bacteria, which would otherwise multiply rapidly in the sugary solution. For the next two weeks, we will investigate the concentration of phosphoric acid in a sample of Coca-Cola Classic. We will compare two experimental techniques.

Spectrophotometric Analysis

We will determine the concentration of phosphoric acid by using spectrophotometry to measure the absorption of the sample and relating it to the concentration of phosphoric acid. We utilized this principle to determine the concentration of iron in a patient's blood sample in Experiment 11.

The Lambert–Beer law states that the absorption of light as it passes through a solution is proportional to the concentration of the absorbing species, the length of the light path, and a fundamental property of the material called the molar absorptivity, ε.

$$A = \varepsilon l c$$

A solution that absorbs light in the visible range is colored. The color observed is dependent upon the wavelengths of light that pass through the solution. This experiment measures the amount of light absorbed and relates it to the concentration of the absorbing substance.

To analyze cola, we must overcome limitations to use this method. First, the molar absorptivity of phosphoric acid is quite small at all wavelengths. We need to produce a colored solution that absorbs in the visible region. The reaction of a phosphate species with an ammonium molybdate/ammonium metavandate mixture produces an intensely colored heteropoly acid, $(NH_4)_3PO_4 \cdot NH_4VO_3 \cdot 16\ MoO_3$. This yellow complex absorbs strongly at 400 nm.

The second problem requires a color correction. Coca-Cola Classic is a dark color and contains light absorbers other than phosphoric acid. You will prepare a color blank by diluting the cola fiftyfold and using this sample as a reference for the cola sample.

Procedure

Place ~100 mL of Coca-Cola Classic in an Erlenmeyer flask. Mark the liquid level with a pen and cover the flask with a watch glass. Heat the sample for ~20 minutes to expel carbon dioxide. After you have stopped heating the sample and it has cooled to room temperature, add distilled water to restore the liquid level to the original mark.

Preparation of Calibration Curve

Record the concentration of the standard phosphate solution. Use a clean 50.00 mL volumetric flask. Rinse the flask with distilled water, and use a volumetric pipet to dispense exactly 0.25 mL of the standard phosphate solution into the flask. Next, add exactly 10.00 mL of the color-enhancing reagent to the flask. Dilute to the 50.00 mL mark with distilled water. Cap the flask, and invert several times to ensure complete mixing. Transfer some of the mixture to a clean, dry, labeled, test tube. Set the solution aside to allow complete color development. Repeat the procedure using 0.50 mL, 0.75 mL, and 1.0 mL of the standard phosphate solution.

Prepare your cola sample by pipeting 1.00 mL of the cooled, decarbonated cola into a 50.00 mL volumetric flask. Add 10.00 mL of the ammonium molybdate/ammonium metavandate color-enhancing solution, and dilute to the mark with distilled water. Cap and invert to mix well. Transfer a portion of this solution to a dry, clean, labeled test tube. Allow the color to fully develop. Prepare your color blank by pipeting 1.00 mL of the cooled, decarbonated cola into a 50.00 mL volumetric flask. Dilute to the mark with distilled water. Mix well, and transfer a portion to another clean, dry, labeled test tube.

The spectrophotometer should be set to a wavelength of 400 nm. Use distilled water as a reference for the phosphate standards, and use the color blank as a reference for the cola sample. Record the absorbance value or read the % transmittance and convert to absorbance. Plot the absorbance versus concentration for the phosphate standards. Apply the Lambert–Beer law to determine the molarity of the phosphoric acid in your diluted cola sample. Calculate the molarity of phosphoric acid in your original cola sample. Determine the percentage of phosphoric acid in Coca-Cola Classic. (Assume that the density of cola is 1.00 g/mL.)

REFERENCE

1. http://www.worldofcoca-cola.com/secret-vault.htm

Name _____ Date _____

Instructor _____

Analysis of Phosphoric Acid in Coca-Cola Classic: Spectrophotometric Analysis

1. Why must you decarbonate the cola? Comment on how carbonation may interfere with the data collection and results.

2. Why will we measure the absorbance of the solutions at 400 nm?

3. What is the purpose of preparing a calibration plot?

SPECTROPHOTOMETRIC DETERMINATION

Part I: Data Table:

Concentration of Phosphate Standard   _____

Volume of Standard	Final Concentration	% Transmittance	Absorbance
0.25 mL	_____	_____	_____
0.50 mL	_____	_____	_____
0.75 mL	_____	_____	_____
1.00 mL	_____	_____	_____
Cola Sample	_____	_____	_____

Part II: Graphical Analysis

Attach a graph of absorbance versus concentration. Indicate the molar absorptivity.

Use the Lambert–Beer's Law equation to calculate the concentration of phosphoric acid in the cola sample. Show all steps of your calculation.

Molarity of H_3PO_4 in original beverage   _____

Percentage of H_3PO_4 in original beverage   _____

Recently, a local pond has been overtaken by algae. Design an experiment to test for phosphate content in the local pond water.

Analysis of Phosphoric Acid in Coca-Cola Classic: pH Titration | 20

pH Titration

Since cola beverages are dark in color, a traditional acid/base titration using an indicator is not suitable, as a color change cannot be used to detect the endpoint. However, pH titrations provide an accurate and convenient experimental technique for measuring dissociation constants for weak acids. We can also perform a pH titration to determine the phosphoric acid concentration in a sample of cola.[1] Phosphoric acid, H_3PO_4, is a triprotic acid, meaning there are three ionizable protons. The reaction showing the dissociation of each of these protons has its own value of K_a, as shown below.

$$H_3PO_4 + H_2O \leftrightarrow H_3O^+ + H_2PO_4^- \qquad\qquad K_{a1} = \frac{[H_3O^+][H_2PO_4^-]}{[H_3PO_4]} \qquad (20.1)$$

$$H_2PO_4^- + H_2O \leftrightarrow H_3O^+ + HPO_4^{2-} \qquad\qquad K_{a2} = \frac{[H_3O^+][H_2PO_4^{2-}]}{[H_2PO_4^-]} \qquad (20.2)$$

$$HPO_4^{2-} + H_2O \leftrightarrow H_3O^+ + PO_4^{3-} \qquad\qquad K_{a3} = \frac{[H_3O^+][HPO_4^{2-}]}{[PO_4^{3-}]} \qquad (20.3)$$

The dissociation of the phosphoric acid takes place in a stepwise fashion, meaning that all H_3PO_4 molecules are transformed to $H_2PO_4^-$ ions before any substantial amount of HPO_4^{2-} is formed. During your titration, you will observe two equivalence points. It is not possible to titrate the third proton of H_3PO_4 successfully. Thus, the value of K_{a3} will not be obtained.

Procedure

Place ~100 mL of Coca-Cola Classic in an Erlenmeyer flask. Mark the liquid level with a pen, and cover the flask with a watch glass. Heat the sample for ~20 minutes to expel carbon dioxide. After you have stopped heating the sample and it has cooled to room temperature, add distilled water to restore the liquid level to the original mark.

Calibrate the pH probe using a pH 7.0 buffer and a pH 4.0 buffer.

Prepare a buret by rinsing first with distilled water and next a small portion of NaOH. Fill the buret with NaOH, and record the exact molarity listed on the bottle. Place a magnetic stir bar in the flask containing the sample. Place the flask on a magnetic stirrer. Carefully immerse the

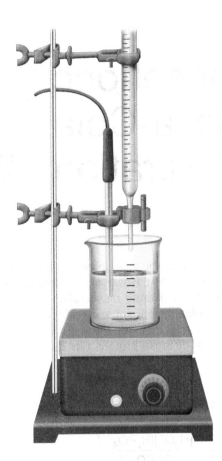

FIGURE 20.1 Do not let the electrode interfere with the spinning of the stir bar.
Adapted from Experiments in General Chemistry Laboratory Manual, by Daphne Norton, Hayden McNeil Publishing.

electrode in the sample, and avoid contact between the tip of the electrode and the spinning stir bar as seen in Figure 20.1. Prepare a data table that records the following three entries: buret reading in mL, volume of base added in mL, and pH.

Read and record the initial pH, and begin the titration by adding 1–2 mL of the standardized NaOH. As you approach the equivalence point, decrease the volume of each increment of base that is added until you are adding only single drops. Continue the titration until well past the second equivalence point.

Data Analysis

Prepare a graph of pH versus volume of base added. Locate the first equivalence point. Combine the molarity of the base with the volume used to reach the equivalence point to find the concentration of phosphoric acid in the beverage.

Find the pH of the halfway point for the first portion of the curve to find a value of K_{a1} for phosphoric acid. Compare this value to the literature value. Now calculate the independent value of K_{a1} by combining the initial pH of the cola with the concentration of phosphoric acid in equation 20.1. Compare this value of K_{a1} with the literature value. Find the percent ionization of the phosphoric acid in the beverage.

Evaluate K_{a2} by finding the pH halfway between the first and second equivalence points. Compare this value to the literature value. Calculate the ionization of $H_2PO_4^-$ in your sample. Calculate an independent value for the concentration of phosphoric acid in the cola based on the volume of base used to reach the second endpoint.

REFERENCE

1. Murphy, Joe. 1983. *Journal of Chemical Education, 60* (5): 420.

Name _____ Date _____

Instructor _____

PRE-LAB QUESTIONS | EXPERIMENT

Analysis of Phosphoric Acid in Coca-Cola Classic: pH Titration | 20

1. Explain the difference between an endpoint and an equivalence point in a titration.

2. Why is it important to calibrate the pH meter?

3. The NaOH solution used during the titration will be standardized by the lab prep staff before it is used in the experiment. Why is this necessary?

Name _____ Date _____

Instructor _____

REPORT SHEET | EXPERIMENT

Analysis of Phosphoric Acid in Coca-Cola Classic: pH Titration | 20

pH Titration

Data Table

Volume of Coca-Cola sample _____

Concentration of NaOH _____

Initial buret reading (NaOH) _____

Volume of NaOH for first equivalence point _____

Molarity of H_3PO_4 _____

pH at halfway point to first equivalence point _____

Experimental K_{a1} _____

Literature value for K_{a1} _____

Value of K_{a1} from initial pH _____

Percent ionization of H_3PO_4 _____

Value of K_{a2} from titration curve _____

Literature value of K_{a2} _____

Percent ionization of $H_2PO_4^-$ _____

Volume of NaOH to reach second equivalence point _____

Molarity of H_3PO_4 _____

Note: You must attach a graph of pH versus volume of NaOH added. The graph must have K_{a1} and K_{a2} labeled. A copy of this graph should be added to your lab notebook.

Overall Comparison

1. Compare the results from the two methods in experiments 19 and 20, spectrophotometric analysis and pH titration.

2. If you could use only one method for analysis, which would you choose and why?

3. Sprite does not contain phosphoric acid. Only sodas that are dark in color contain this acid. Clear sodas such as Sprite contain citric acid, sugar, and carbonated water. Design a simple experiment to determine the concentration of citric acid in Sprite. Provide a detailed procedure.

Borax Solubility: Investigating the Relationship between Thermodynamics and Equilibrium | 21

Borax, also known as sodium tetraborate, is a naturally occurring mineral. Powdered borax consists of soft crystals that may be dissolved in water. Borax was first discovered in dry lake beds in Tibet as traders traveled the Silk Road.[1] The compound was later mined in California and Nevada and marketed in the late 19th century for its many household uses. Borax is an active ingredient in some detergents and most ant killers. It is also used as a water softener.

Borax is slightly soluble in cold water but readily soluble in hot water. This allows its application as a laundry detergent.

The dissolution of borax in water is shown by the following equation.

$$Na_2[B_4O_5(OH)_4] \cdot 8\ H_2O(s) \rightleftharpoons 2\ Na^+(aq) + B_4O_5(OH)_4{}^{2-}(aq) + 8\ H_2O(l)$$

Solubility product constants are used to describe saturated solutions of ionic compounds with relatively low solubility. A saturated solution is in a state of dynamic equilibrium between the dissolved, dissociated, ionic compound and the undissolved solid. The solubility product constant, K_{sp}, for this equation is related to the ion concentrations in the saturated solution as follows. Recall that K_{sp} is dependent on temperature.

$$K_{sp} = [Na^+]^2[B_4O_5(OH)_4{}^{2-}]$$

Let x signify the solubility of borax. In a saturated solution, the concentration of tetraborate ion, $B_4O_5(OH)_4{}^{2-}$, is equal to x, and the concentration of sodium ion is equal to 2x. The solubility product expression then becomes as follows.

$$K_{sp} = [2x]^2[x] = 4x^3$$

The tetraborate ion formed in solution is a weak base and reacts with water according to the following reaction:

$$B_4O_5(OH)_4{}^{2-}(aq) + 5\ H_2O(l) \rightleftharpoons 4\ H_3BO_3(aq) + 2\ OH^-(aq)$$

The concentration of $B_4O_5(OH)_4{}^{2-}$ ion in a saturated solution of borax may be determined by titration with a strong acid according to the following equation:

$$B_4O_5(OH)_4{}^{2-}(aq) + 2\ H_2O(aq) + 3\ H_2O(l) \rightleftharpoons 4\ H_3BO_3(aq)$$

You can measure the amount of borate ion in solution by titrating with HCl. This allows you to determine the solubility of borax at various temperatures. A K_{sp} value for each temperature will be calculated using the equation $K_{sp} = 4x^3$.

Thermodynamics and Equilibrium

During the experiment, you will observe the solubility behavior of borax at varying temperatures. You encounter this in your daily life when you notice that sugar dissolves more quickly and to a greater extent in your hot coffee than in a cold beverage. Other compounds such as calcium carbonate are less soluble in warm water than cold water.

Your first step is to determine the K_{sp} value. Next, you will be able to use the relationship between the equilibrium constant and the change in standard free energy, $\Delta G°$, to derive thermodynamic quantities.

Chemical reactions involve transformations in energy as well as transformations in matter. Thermodynamics is the study of the relationship between heat and other forms of energy. The standard enthalpy change, $\Delta H°$, indicates the amount of heat absorbed or released as a process occurs at constant pressure and under standard conditions. The standard entropy change, $\Delta S°$, indicates the change in disorder in the system as a process occurs under standard conditions. In general, a spontaneous process is favored by an increase in disorder and a release of heat. However, in order to determine whether or not a process will occur in the forward direction under specific conditions, *both* $\Delta H°$ and $\Delta S°$ must be considered. The enthalpy change and the entropy change for a process are combined in the thermodynamic quantity known as standard free energy change or Gibbs free energy, $\Delta G°$.

$$\Delta G° = \Delta H° - T\Delta S°$$

If $\Delta G°$ is negative, the process will occur in the forward direction at the specified temperature. If $\Delta G°$ is positive, the process will not occur in the forward direction at that temperature; it will be spontaneous in the reverse direction. If $\Delta G°$ is zero, the process is at equilibrium, and there will be no net change in either direction at that temperature. The relationship between the standard free energy change and the equilibrium constant, K, is shown in the following equation.

$$\Delta G° = -RT\ln K$$

Note that if K is greater than one, then $\Delta G°$ is negative, indicating that the process will proceed in the forward direction under the standard conditions.

Combination of the relationships $\Delta G° = \Delta H° - T\Delta S°$ and $\Delta G° = -RT\ln K$ results in the following equation.

$$\Delta H° - T\Delta S° = -RT\ln K$$

Note: Both $\Delta H°$ and $\Delta S°$ are temperature dependent. However, for the purpose of this exercise, it may be assumed that they do not change substantially with temperature. This introduces a degree of error into the results.

Rearranging this equation gives the following.

$$\ln K = \frac{-\Delta H°}{RT} + \frac{\Delta S°}{R}$$

A plot of lnK versus 1/T should give a straight line with slope equal to $-\Delta H°/R$. Theoretically, the y-intercept of the plot, where x = 0, should be equal to $\Delta S°/R$. However, x is 1/T, and there is no y-value for which 1/T = 0. Thus, $\Delta S°$ may not be determined directly from the plot.

Procedure

Fill an 800 mL beaker about three-quarters full with distilled water. Place the beaker on a hot plate, and put a large-mouthed graduated pipet into the water.

You will do a series of five titrations. Label five clean 125 mL Erlenmeyer flasks as 33 °C, 36 °C, 39 °C, 42 °C, and 45 °C. Add approximately 5 mL of distilled water to each flask, and place them out of the way.

Next, you will prepare a saturated borax solution. Measure ~18 grams of borax into a 100 mL beaker, and add 70 mL of distilled water. Place the mixture on the hot plate, and stir occasionally. Monitor the temperature closely. When the temperature reaches 65 °C, look for solid salt in the beaker. If you observe undissolved salt, you have formed a saturated solution. (If all of the borax is completely dissolved at 65 °C, add more borax to the beaker while continuing to heat until some of the salt remains undissolved.) Use beaker tongs to remove the saturated solution from the hot plate. Allow the solution to cool.

Continue to heat the beaker of water with the pipet.

When the borax solution has cooled to ~45 °C, use the large-mouthed pipet to remove a 5 mL aliquot and transfer to the appropriately labeled flask. Try not to allow any undissolved borax to be transferred. Use the same pipet to transfer 5 mL of the warm distilled water into the flask. This rinses the pipet and prevents the tips from becoming clogged with borax. Record the actual solution temperature to 0.1 °C. Repeat this procedure for each of the five temperatures.

After all of the aliquots have been collected, add an additional 20 mL of distilled water as well as 4 drops of bromcresol green to each flask. The solutions will appear light blue in color.

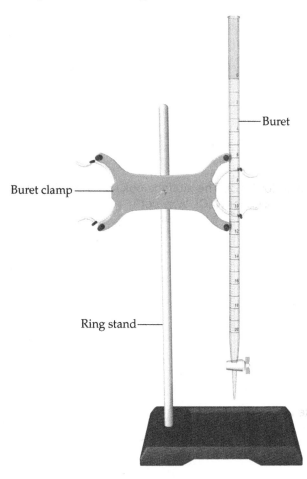

FIGURE 21.1 Setting up the buret.
From Tro, Laboratory Experiments for Chemistry: A Molecular Approach 3e. © Pearson Education, Inc.

Set up a buret as shown in Figure 21.1. Rinse the buret several times with distilled water, and then rinse with a small amount of HCl solution. Fill the buret with the HCl solution, and record the concentration of the HCl solution. Record the initial buret reading, and slowly add the acid dropwise. The endpoint is observed when the solution turns from blue to pale yellow. Titrate each borax solution, recording the initial and final buret readings.

REFERENCE

1. http://www.thechemco.com/chemical/borax/

PRE-LAB QUESTIONS | EXPERIMENT

Borax Solubility: Investigating the Relationship between Thermodynamics and Equilibrium | 21

1. Look up the K_{sp} values for the following compounds: lead (II) sulfate, aluminum hydroxide, silver chloride, silver bromide, and copper sulfide. Which one is most soluble? least soluble? Explain.

2. A student is curious about the K_{sp} value for NaCl. The student looks up the value in the appendix of his textbook but can't find a value for NaCl. Why not?

3. Draw the structure for the tetraborate ion.

Name _____ Date _____

Instructor _____

REPORT SHEET | EXPERIMENT

Borax Solubility: Investigating the Relationship between Thermodynamics and Equilibrium | 21

Complete the data table for each temperature studied.

Beaker Label	33 °C	36 °C	39 °C	42 °C	45 °C
Temperature reading (°C)					
Absolute temperature (K)					
1/T					
Final buret reading					
Initial buret reading					
Volume of HCl					
Moles of H^+					
Moles of $B_4O_5(OH)_4^{2-}$					
Molarity of $B_4O_5(OH)_4^{2-}$					
K_{sp}					
$\ln(K_{sp})$					

Graph ln K_{sp} versus 1/T. Attach your graph.

Determine $\Delta H°$ from the slope of the line.

Determine $\Delta S°$ for one of the temperature measurements. Show your work.

Determine $\Delta G°$ at 25 °C. Show your work.

Electrochemical Preparation of Nickel Nanowires | 22

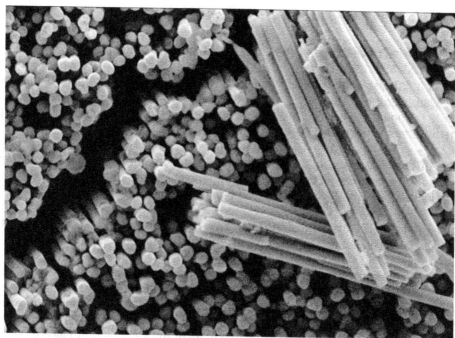

Daphne Norton

Nanotechnology is the science of building devices at the atomic or molecular level. The nanoscale represents materials developed on the nanometer level (10^{-9} m). Advances in nanotechnology lead to smaller, lighter, and more efficient technology that functions at the atomic level. Basically, scientists are manipulating individual atoms to create materials with desired properties. By exploiting materials at the nanometer scale, biological, chemical, and physical properties can be controlled. Scientists must use advanced microscopy techniques to develop and characterize these new materials. To better understand the scale at which these nanoscientists are working, consider that the diameter of a human hair is approximately 100,000 nm. This is 100,000 times larger than a single-walled carbon nanotube.[1]

In this experiment, we will synthesize nickel nanowires.[2,3] We will use electrolysis to grow nickel wires in the pores of an alumina filter. The diameter of these wires is on the nanoscale. A technique called electrodeposition will be used to selectively grow the wires. Electrodeposition involves passing an electric current through an electrochemical cell. The diameter of the wire we grow will be the same diameter as the pore in the filter. We can control the diameter of the wire by choosing a filter with the desired pore size. The micrograph shown above shows nickel nanowires prepared by this electrochemical process.

Sometimes reactions are spontaneous. They occur naturally without external forces. In an electrolytic reaction, an external current is supplied from an outside source such as a battery.

The cells contain **electrodes** where oxidation and reduction reactions take place. Oxidation occurs at the **anode**, and reduction occurs at the **cathode**. These reactions take place in a medium called an **electrolyte**.

In our reaction, our anode, a nickel wire, will undergo oxidation to form Ni^{2+} cations. A corresponding reduction will take place at the cathode. The Ni^{2+} ions will be reduced back to elemental nickel.

At first glance, this seems pointless. The reduction is undoing what has just been achieved through the oxidation. However, the reduction at the cathode involves selective deposition. We are not simply reducing the ions. We are controlling the formation of wires by depositing the ions into a porous membrane or alumina filter. By controlling the size of the pores of the membrane, we can control the diameter of the wire that forms through electrochemical deposition.

Procedure

Collect a piece of nickel wire to use as the anode. Next, prepare the cathode.

The cathode is a flat disc composed of alumina (Al_2O_3.) These discs are very brittle and will easily crack. Use tweezers to delicately remove one Anodisc filter from the storage box. Note which side was facing up in the box. The discs are supported by a polystyrene (plastic) support ring. Use the tweezers to hold the disc by the support ring.

The pore size of this filter disc is 0.02 micrometer or 20 nanometers. This will be the diameter of our deposited wires. In order for the nickel ions to deposit into the pores of the filter, we must make it a conductive medium. We will coat it with a gallium-indium, GaIn, alloy.

Use a cotton swab to apply a thin coat of the GaIn alloy to the surface of the disc. You must coat the side that was facing up in the box. The polystyrene ring looks wider from this side.

Now, you will assemble your cathode. Take a piece of copper sheet, and wipe it clean with acetone and let it dry. Place the filter shiny face down on the sheet of copper. (The coated side of the disc must face the copper metal.) Carefully tape the disc in place without touching the surface of the filter. Tape should only be in contact with the support ring. Next, cover the entire back and sides of the copper sheet with electrical tape so that none of the electrolyte solution will come in contact with the copper electrode.

Assemble your electrochemical cell using the diagram in Figure 22.1. Pour the nickel electrolyte solution in the beaker containing the electrodes. The positive lead should be connected to the anode using an alligator clip. The anode is your nickel wire. The negative lead should be connected to the copper sheet using the alligator clip because the copper sheet is the cathode.

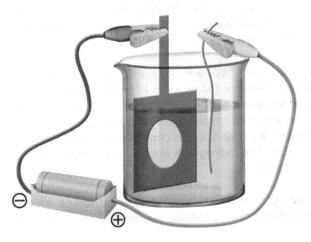

FIGURE 22.1 The nickel wire connects to the positive electrode, and the copper plate connects to the cathode or negative electrode.
Adapted from Experiments in General Chemistry Laboratory Manual, Hayden McNeil Publishing, used by permission.

Allow the electrolysis to run for 15–20 minutes. When the time is up, remove the clamps and rinse the copper sheet with distilled water. Place the copper sheet in a beaker of acetone. Let this sit for 5–10 minutes. The acetone will dissolve the tape residue.

Carefully remove the filter disc and place it on a microscope slide with the shiny side up. Dip a cotton swab in nitric acid, and remove the GaIn coating. Do this step in a hood, and place used swabs in the appropriate waste container. Once all of the GaIn coating is removed, place the filter disc in a small beaker with 15 mL of 6 M NaOH. *(Caution: Sodium hydroxide is caustic. The NaOH should be added in a fume hood.)* Allow the disc to sit in the solution for 10 minutes as the base dissolves the porous membrane. Use tweezers to remove the plastic support ring.

Place a strong magnet under the beaker. Decant the NaOH while magnetically holding the nanowires in the bottom of the beaker. Rinse with distilled water, again holding the nanowires in place. Suspend the nanowires in 15–20 mL of distilled water, and evaluate using the magnet. Place the magnet 15–20 cm away from the beaker. What happens as you move the magnet closer to the wires? Turn the magnet 90 degrees. What happens now?

REFERENCES

1. http://www.nano.gov/nanotech-101/what/nano-size
2. A. K. Bentley, M. Farhoud, A. B. Ellis, G. C. Lisensky, Anne-Marie Nickel, and W. C. Crone. 2005. "Template Synthesis and Magnetic Manipulation of Nickel Nanowires," *Journal of Chemical Education*, 82, 765–768.
3. http://education.mrsec.wisc.edu/287.htm

PRE-LAB QUESTIONS | EXPERIMENT

Electrochemical Preparation of Nickel Nanowires | 22

1. Complete the table below. It may be helpful to use scientific notation.

red blood cell	2.5 micrometers	nanometers
water molecule	280 picometers	nanometers
height of Mount Everest	8.8 kilometers	nanometers
diameter of a proton	1 femtometer	nanometers
grain of salt	500 micrometers	nanometers
Boeing 747 jet airplane	65 meters	nanometers
length of ant	4 millimeters	nanometers
length of O—H bond in H_2O	0.96 angstroms	nanometers
height of giraffe	6 meters	nanometers
diameter of Earth	12,700 kilometers	nanometers
infrared wavelength	15 micrometers	nanometers
length of matchstick	5 centimeters	nanometers

2. Define electrolytic cell.

3. Define cathode and anode.

4. Why must you tape the copper electrode with electrical tape?

5. Give two examples of materials or procedures that have been enhanced by advances in nanotechnology. Explain the impact. Cite your references.

6. Cite differences between the properties of bulk Ni and nickel nanowires.

Name _____ Date _____

Instructor _____

REPORT SHEET | EXPERIMENT

Electrochemical Preparation
of Nickel Nanowires | 22

1. In our experimental setup, what serves as the cathode? the anode?

2. Write equations to represent what is happening at the anode and the cathode.

3. The nickel solution can be reused for this experiment. Why doesn't the concentration of nickel in solution change during the electrolysis?

4. Write a reaction to explain the dissolution of alumina in NaOH.

5. If your goal had been to quantify the amount of Ni that forms nanowires, what would be an easy way to estimate this value? Explain your procedure.

6. Cite at least two possible applications of nickel nanowires.

Synthesis of $K_3Fe(C_2O_4)_3 \cdot 3\,H_2O$

Coordination compounds consist of a transition metal ion that forms when ligands bind to a central metal atom. A *ligand* is a neutral molecule or ion that has a lone pair of electrons that can be used to form a bond with the metal ion. When a ligand donates a lone pair, it functions as a Lewis base. The metal acts as a Lewis acid. Common ligands include Cl^-, Br^-, I^-, H_2O, CO, NH_3, and CN^-.

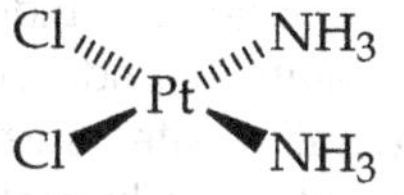

FIGURE 23.1 Cisplatin is the first in a series of platinum-containing anticancer drugs.

A very famous coordination compound is *cis*-diamminedichloroplatinum(II), commonly called cisplatin. This cancer-fighting compound, shown in Figure 23.1, consists of a platinum metal center bound to two chloride ligands and two amine ligands. The metal has a coordination number of 4, and the geometry is square planar. Upon injection into the human body, one of the chloride ligands is slowly displaced by water to form $[PtCl(H_2O)(NH_3)_2]^+$. The aqua (water) ligand is easily displaced, allowing the platinum atom to bind to bases in DNA. DNA crosslinking can occur, which ultimately triggers apoptosis or programmed cell death.[1]

Further advances in medicinal chemistry led to derivatives of this platinum-containing anticancer treatment. Sometimes coordination compounds contain ligands with more than one electron pair available for binding. This is the case in oxaliplatin. Oxaliplatin features two bidentate ligands binding to the Pt center. A bidentate ligand has two coordination sites. The two monodentate amine ligands in cisplatin were replaced with the bidentate ligand, 1,2-diaminocyclohexane. The compound, as shown in Figure 23.2, also features a bidentate oxalate group.[2] The coordination number for Pt is still 4, but two ligands are coordinated to the metal instead of four.

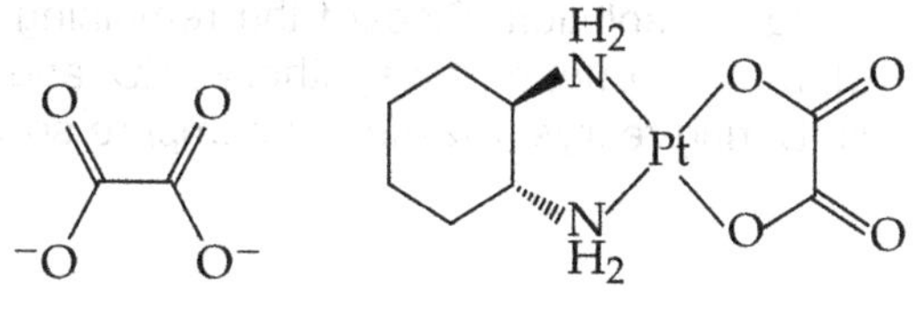

(a)   **(b)**

FIGURE 23.2 (a) A free oxalate ligand. (b) Pt shown bound to an oxalate ligand in oxaliplatin.

Other ligands have even more bonding sites. These are called polydentate ligands. Ethylenediaminetetraacetic acid or EDTA is a classic example of a polydentate ligand. It can completely surround a metal forming six coordinate covalent bonds, as shown in Figure 23.3. EDTA is used to sequester metals in the treatment of lead and mercury poisoning. When bound to iron, EDTA is used to "fortify" grain-based cereals. Complexes containing polydentate ligands are very stable because it would require breaking multiple bonds to displace the ligand.[3]

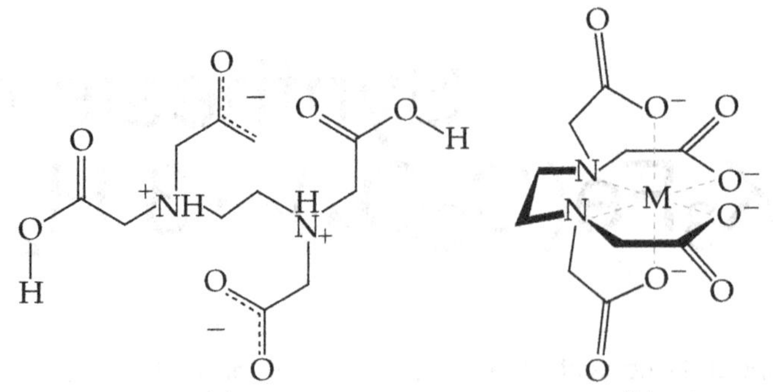

FIGURE 23.3 (a) Free EDTA ligand. (b) EDTA coordinating to a metal.

You will synthesize a coordination compound with bidentate ligands surrounding an iron atom. $K_3Fe(C_2O_4)_3 \cdot 3\ H_2O$ contains three oxalate ligands coordinated to the central iron (III) atom. Three oxalate ligands can fill six coordination sites surrounding the iron. Water is not acting as a ligand. The compound contains three waters of hydration. Three potassium ions are needed to balance the charge on $Fe(C_2O_4)_3{}^{3-}$.

Three equivalents of potassium oxalate can be reacted with iron (III) chloride to produce potassium trioxalatoferrate (III) trihydrate.

$$FeCl_3 + 3\ K_2C_2O_4 + 3\ H_2O \rightarrow K_3Fe(C_2O_4)_3 \cdot 3\ H_2O + 3\ KCl$$

Procedure

Add ~8.0 grams of 3 $K_2C_2O_4 \cdot H_2O$ to a clean 50 mL beaker. Dissolve the solid in ~15 mL of water by placing the beaker on a wire gauze atop a ring stand and heating the solution over a Bunsen burner. The setup is shown in Figure 23.4.

Use beaker tongs to carefully remove the beaker and add 5.0 mL of $FeCl_3$ (0.40 g/mL) to the solution while it is still hot. A bright green solution will form.

Prepare an ice bath by placing the smaller beaker containing the solution inside a 250 mL beaker filled with ice. Stir the solution. A green precipitate should form as the solution cools. Allow the solution to cool at least 10 minutes to ensure complete precipitation.

Collect the solid by decanting the solution. Discard the remaining solution. The solid product will most likely contain KCl, a by-product of the synthesis. KCl and other impurities can be removed via *recrystallization*. During recrystallization, the impure solid compound is dissolved in

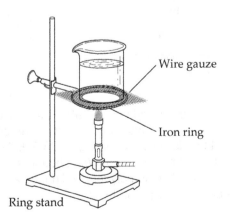

Wire gauze

Iron ring

Ring stand

FIGURE 23.4 Carefully place the beaker on the center of the wire gauze square. The ring should be securely fastened to the ring stand.
From Timberlake, Laboratory Manual for General, Organic, and Biological Chemistry, 3E. © Pearson Education, Inc.

a minimal amount of solvent and then allowed to slowly crystallize as the solution cools. As the compound crystallizes from the solution, the impurities remain in solution and are excluded from the growing crystal lattice, thus resulting in a pure solid.[4]

Dissolve the solid product in 15 mL of deionized water until a clear green solution results. It may be necessary to heat the solution gently. Place the solution in an ice bath and cool for 10–20 minutes. The product may be collected by vacuum filtration. Rinse the collected product with a small (5 mL) portion of acetone. Transfer your product to a watch glass labeled with your initials.

Waste Disposal

Discard all solutions containing iron in the appropriate waste bottle in the hood.

REFERENCES

1. Trzaska, Stephen. June 20, 2005. "Cisplatin." *C&EN News*, 83 (25).
2. Wheate, Nial J., Walker, Shonagh, Craig, Gemma E., and Rabbab Oun. "The Status of Platinum Anticancer Drugs in the Clinic and in Clinical Trials." *Dalton Transactions*, 2010, 39: 8113–8127.
3. http://umm.edu/health/medical/altmed/supplement/ethylenediaminetetraacetic-acid
4. http://www.erowid.org/archive/rhodium/chemistry/equipment/recrystallization.html

PRE-LAB QUESTIONS | EXPERIMENT

Synthesis of $K_3Fe(C_2O_4)_3 \cdot 3\,H_2O$ | 23

1. A reaction of 7.8 mL of iron(III) chloride solution (0.40 g/mL) and 8.64 g of $K_2C_2O_4 \cdot H_2O$ produced 6.07 g of $K_3Fe(C_2O_4)_3 \cdot 3\,H_2O$. Determine the limiting reagent, and calculate the percent yield of the synthesis.

2. Draw the structure of $K_3Fe(C_2O_4)_3 \cdot 3\,H_2O$.

3. Explain the difference between precipitation and recrystallization.

Name _____ Date _____

Instructor _____

REPORT SHEET | EXPERIMENT

Synthesis of $K_3Fe(C_2O_4)_3 \cdot 3\,H_2O$ | 23

Experimental Data

Volume of $FeCl_3$ solution   _____

Mass of $K_2C_2O_4 \cdot H_2O$ _____

Mass of product   _____

Calculations

Moles of $FeCl_3$ _____

Moles of $K_2C_2O_4 \cdot H_2O$ _____

Moles of product ($K_3Fe(C_2O_4)_3 \cdot 3\,H_2O$)   _____

Limiting reagent   _____

Percent yield   _____

Comment on possible impurities in your product:

Show all work for determining the limiting reagent and percent yield.

Why did you rinse your product with acetone instead of water?

Synthesis of $K_3[Fe(C_2O_4)_3]\cdot 3H_2O$ 23

Experimental Data

Calculations

Analysis of Oxalate in $K_3Fe(C_2O_4)_3 \cdot 3\ H_2O$ | 24

In Experiment 23, last week you synthesized a metal oxalate complex, potassium trioxalatoferrate (III) trihydrate. For the experiment, this week you will determine the percent composition of oxalate in this compound. The percent oxalate will be determined by titrating a solution of $K_3Fe(C_2O_4)_3 \cdot 3\ H_2O$ with a standardized solution of $KMnO_4$, potassium permanganate.

Under acidic conditions, $KMnO_4$ oxidizes $C_2O_4^{2-}$ to produce CO_2 gas, as shown in this equation:

$$16\ H^+ + 5\ C_2O_4^{2-} + 2\ MnO_4^- \rightarrow 10\ CO_2 + 2\ Mn^{2+} + 8\ H_2O$$

A color indicator cannot be used for this titration because the titrant, potassium permanganate, is a deep purple color. The volume and molarity of the titrant can be used to determine the number of moles of oxalate in the sample. The reaction between permanganate ions and oxalate ions is very slow. The reaction must be heated to 60 °C and kept warm throughout the titration.

Standardization of $KMnO_4$

The molarity of the titrant, $KMnO_4$, must be known with great accuracy to be used in the calculation of oxalate content. The $KMnO_4$ solution can be standardized using sodium oxalate, $Na_2C_2O_4$, as a primary standard. $Na_2C_2O_4$ is a solid that can be easily measured. It is stable and non-hygroscopic and can be produced with a high level of purity. By measuring the mass of the solid sodium oxalate on an analytical balance, you can determine the number of moles of sodium oxalate with a high level of accuracy. Thus, you can determine the exact concentration of the $KMnO_4$ solution.

The standardized solution of $KMnO_4$ can then be used to titrate $K_3Fe(C_2O_4)_3 \cdot 3\ H_2O$ and determine the percent oxalate in the sample.

Procedure

First, measure approximately 0.15 g of $Na_2C_2O_4$ and add it to your clean, dry 250 mL Erlenmeyer flask. Add 100 mL of 0.5 M H_2SO_4 and dissolve. Next add 1 mL of 85% phosphoric acid, H_3PO_4. Carefully heat the solution to 60 °C using a hotplate. *(Caution: H_2SO_4 is a strong acid. If any acid comes in contact with your skin, notify your instructor immediately and rinse the area with water.)*

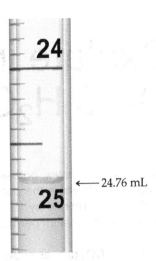

FIGURE 24.1 The buret is marked to the tenth place. Therefore, you can estimate the value to the hundredth place. The buret in the figure displays a final volume of 24.76 mL.
From Tro, Laboratory Manual for Chemistry: A Molecular Approach 3e. © Pearson Education, Inc.

← 24.76 mL

Prepare your buret by rinsing with distilled water followed by a small portion of 0.01 M $KMnO_4$. Fill the buret with 50.00 mL of 0.01 M $KMnO_4$, and titrate the solution until a faint pink color exists for at least 30 seconds. Your initial and final buret readings should be recorded to the hundredth place, as shown in Figure 24.1. Repeat the titration. You should have two trials. Use these data to calculate the exact molarity of the $KMnO_4$ solution.

Next, you will weigh approximately 0.3 g of $K_3Fe(C_2O_4)_3 \cdot 3\,H_2O$ (to the nearest 0.1 mg) and add it to your clean, dry 250 mL Erlenmeyer flask. Add 100 mL of 0.5 M H_2SO_4 and dissolve. Next add 1 mL of 85% phosphoric acid, H_3PO_4. Carefully heat the solution to 60 °C using a hotplate. Titrate your sample using the potassium permanganate solution. Repeat the titration. Use these data to determine the number of moles of oxalate present in your sample. Then, you can calculate the percent oxalate in your sample of $K_3Fe(C_2O_4)_3 \cdot 3\,H_2O$.

PRE-LAB QUESTIONS | EXPERIMENT

Analysis of Oxalate in K₃Fe(C₂O₄)₃ · 3 H₂O | 24

Analysis of Oxalate in $K_3Fe(C_2O_4)_3 \cdot 3\,H_2O$ | **24**

1. What is a primary standard?

2. The equation below represents the titration that will be completed in Experiment 24. It is sometimes called a redox titration because an oxidation–reduction reaction occurs. What is the reducing agent? The oxidizing reagent? Explain your answers.

$$16\,H^+ + 5\,C_2O_4{}^{2-} + 2\,MnO_4{}^- \rightarrow 10\,CO_2 + 2\,Mn^{2+} + 8\,H_2O$$

3. In solution, the ferrioxalate complex undergoes photoreduction. The complex absorbs a photon of light and decomposes to form $Fe(C_2O_4)_2{}^{2-}$ and CO_2. What are the initial and final oxidation states of the iron during this process?

EXPERIMENTAL DATA

Standardization of $KMnO_4$	Trial 1	Trial 2
Mass of $Na_2C_2O_4$	_____	_____
Moles of $Na_2C_2O_4$	_____	_____
Final buret reading ($KMnO_4$)	_____	_____
Initial buret reading ($KMnO_4$)	_____	_____
Volume of $KMnO_4$	_____	_____
Molarity of $KMnO_4$	_____	_____

Average molarity of $KMnO_4$ _____

Show molarity calculations for one trial. Include proper units and significant figures.

Oxalate Determination in K$_3$Fe(C$_2$O$_4$)$_3 \cdot$ 3 H$_2$O

	Trial 1	Trial 2
Mass of K$_3$Fe(C$_2$O$_4$)$_3 \cdot$ 3 H$_2$O	_____	_____
Final buret reading (KMnO$_4$)	_____	_____
Initial buret reading (KMnO$_4$)	_____	_____
Volume of KMnO$_4$	_____	_____
% oxalate in K$_3$Fe(C$_2$O$_4$)$_3 \cdot$ 3 H$_2$O	_____	_____

Average % oxalate in K$_3$Fe(C$_2$O$_4$)$_3 \cdot$ 3 H$_2$O   _____

Show % oxalate calculations for one trial. Include proper units and significant figures.

Questions

Oxalic acid is present in spinach, chard, rhubard, kale, and some other leafy green vegetables. Design an experiment to quantify the oxalic acid content in a one-cup helping of spinach. Give specific experimental details.